Conceptual Form

Conceptual Form

The Hidden Structure of Commonsense Concepts

Sandeep Prasada

The MIT Press
Cambridge, Massachusetts
London, England

The MIT Press
Massachusetts Institute of Technology
77 Massachusetts Avenue
Cambridge, MA 02139
mitpress.mit.edu

The MIT Press would like to thank the anonymous peer reviewers who provided comments on drafts of this book. The generous work of academic experts is essential for establishing the authority and quality of our publications. We acknowledge with gratitude the contributions of these otherwise uncredited readers.

This book was set in Stone Serif and Stone Sans by Westchester Publishing Services. Printed and bound in the United States of America.

Library of Congress Cataloging-in-Publication Data

Names: Prasada, Sandeep author
Title: Conceptual form : the hidden structure of commonsense concepts / Sandeep Prasada.
Description: Cambridge, Massachusetts : The MIT Press, [2026] | Includes bibliographical references and index.
Identifiers: LCCN 2025041944 (print) | LCCN 2025041945 (ebook) | ISBN 9780262053136 paperback | ISBN 9780262053143 pdf | ISBN 9780262053150 epub
Subjects: LCSH: Concepts
Classification: LCC BF311 .P7338 2026 (print) | LCC BF311 (ebook)
LC record available at https://lccn.loc.gov/2025041944
LC ebook record available at https://lccn.loc.gov/2025041945

EU Authorised Representative: Easy Access System Europe, Mustamäe tee 50, 10621 Tallinn, Estonia | Email: gpsr.requests@easproject.com

It is only while our attention is turned toward a different object that we are able to perceive, incidentally, the mental processes that are directed towards that object.
—Franz Brentano (1982, 30)

Perhaps here Brentano is echoing a remark of Aristotle's in the *Metaphysics* (book 12, chap. 9), "Knowledge and perception and opinion and understanding always have something else as their object, and themselves only *by the way*" (1074b, 35–36).
—John Macnamara (1999, 215)

We don't get a full definition but we do get structure, so there is something to look at between atomism and definitions.
—Noam Chomsky (2009, 296)

Contents

Preface and Guide to the Book ix
Acknowledgments xiii

I Introduction

1 **Situating the Study of Conceptual Form** 3
2 **The Perspectival Nature of Commonsense Concepts** 21

II Conceptual Form of Kind Representations

3 **The Instance-of-Kind Perspective and Limits of Existing Approaches** 37
4 **Mechanisms for Thinking About Kinds and Instances of Kinds** 51
5 **Generating Perspectives** 79

III Linguistic Applications

6 **New Perspectives on Generics** 109
7 **Explaining Systematic Polysemy** 147
8 **Metaphorical Perspectives** 167

IV Acquisition

9 **Acquiring Perspectives** 193

V Conclusions

10 **The Nature of Conceptual Knowledge** 213
11 **New Horizons for the Study of Conceptual Form** 243
12 **A Bird's-Eye Perspective on Conceptual Form** 265

Notes 269
References 277
Index 297

Preface and Guide to the Book

This book argues that lexically expressible concepts such as DOG possess a formal structure—a *conceptual form*—that provides perspectives for thinking, talking, and reasoning about things and their properties. These perspectives are not a property of the things that the concept represents. Rather, they are "mental lenses," or ways of thinking and talking about things and their properties. The conceptual form functions as a mental lens by providing instructions for how to think, talk, and reason about the things that a concept represents. As our concepts direct our attention to what they represent, we are generally unaware of conceptual form except incidentally, in the implications that it has for how we think about the thing that we are attending to. Existing theories of conceptual representation are focused on conceptual content and do not provide a theory of conceptual form or an explanation of the plethora of linguistic and nonlinguistic phenomena that depend on it.

After arguing for the need for a theory of conceptual form in chapters 1–3, this book develops a fragment of the theory of conceptual form for representing kind concepts in chapters 4 and 5. Lexically expressible kind concepts display a wide variety of intricate and abstract perspectives that can overlap and cross-cut each other; however, they do not vary in arbitrary ways. The perspectives provided by various classes of kind concepts vary in terms of a limited number of formal components and how they are realized. As such, it appears that our conceptual systems are partly constituted by a generative system that yields the constrained variety of conceptual forms displayed by lexically expressible kind concepts given relevant circumstances.

Chapters 6, 7, and 8 show how the conceptual forms of kind concepts can account for a wide variety of linguistic and nonlinguistic phenomena pertaining to generic knowledge and generic language, the systematic polysemy of kind terms, and the interpretation of metaphors. And chapter 9 explicates the acquisition problems that the theory of conceptual form poses for both empiricist and nativist approaches. It then proposes a conceptual bootstrapping solution based on core knowledge systems to the main problem faced by a nativist approach to the acquisition of conceptual form.

The theory of kind concepts that emerges from this book is one in which the formal structure of kind concepts provides the means for representing the perspectives that we take when thinking and talking about kinds of things, as well as the means for representing and acquiring *principles* for thinking about kinds, their instances, and their subkinds. This stands in stark contrast to most other work on concepts, which assumes that we acquire facts and generalizations rather than principles. Chapter 10 discusses key differences between the theory of conceptual form and other work on conceptual representation. The theory of conceptual form is a necessary complement to existing work on concepts, which has not considered the formal dimension of commonsense concepts. Chapter 11 presents promising new avenues for developing the theory of conceptual form. Finally, chapter 12 provides some take-home messages via a bird's-eye perspective on the theory of conceptual form and what it affords.

Because this book is intended to be of interest to psychologists, linguists, and philosophers, among others, I have attempted to present the theory and its implications in a manner that will be accessible to a broad audience. This has meant that I have kept descriptions of other approaches to a minimum, describing them only to the extent needed to situate the theory of conceptual form and to show how it can explain a range of phenomena. The goal of this book is not to provide comprehensive reviews of the many important bodies of literature that the book touches on. The aim is to clearly articulate a dimension of commonsense concepts that has not been studied in previous work and to show how it can explain a range of phenomena.

Although it is best to read the chapters in order, the chapters have been written to be relatively free-standing. As a result, key ideas appear multiple times in different contexts. My hope is that the redundancy helps readers internalize those ideas and see their value in explaining a range of linguistic and nonlinguistic phenomena.

Finally, I want to note that many of the ideas and proposals developed in this book have been inspired by the idea that the formal dimension of commonsense conception may, in important respects, be similar to the formal structure of an Aristotelian metaphysics. I mention these connections as they arise in the course of the book. I do this for three reasons. First, I want to acknowledge the source of inspiration for some of these ideas, even if they have been developed and put to use in ways that would be completely foreign to Aristotle. Second, knowing the source of these ideas may be useful in revising, elaborating, and extending the current approach to account for a larger range of conceptual phenomena. Third, in chapter 11, I suggest that the proposals developed in this book may suggest some new ways of understanding certain aspects of Aristotle's work. That said, it is important to keep in mind that the goal of this book is to make a contribution to cognitive psychology by advancing our understanding of the formal dimension of concepts and how it enters into the acquisition and use of conceptual knowledge to think, speak, and reason. The proposals should be judged based on how well they accomplish this goal rather than their fidelity to what Aristotle may have said (rarely an uncontroversial matter).

Acknowledgments

An unanticipated pleasure of writing this book is the opportunity to recognize the contributions and support of my many teachers, mentors, collaborators, colleagues, students, editors, friends, and family.

I came across the idea that lexically expressible concepts provide intricate and nonobvious perspectives for thinking and talking via Noam Chomsky, who has provided compelling examples of such perspectives on many occasions (usually to argue against externalist referential theories of meaning). Although Chomsky has not provided any theory of the phenomenon, he has suggested the relevance of translating Aristotle's ideas concerning the nature of things into a theory of lexical meaning along the lines of Julius Moravcsik's work on the topic. That idea has suffused my work on conceptual representation for the past three decades and is on display in this book. John Macnamara introduced me to the idea that in learning a count noun, we acquire principles of application, identity, and individuation. The genesis of this book was the idea that those principles are a core part of the perspectives provided by our most basic concepts.

Other key ideas that have guided the development of the theory of conceptual form include Jerry Fodor's idea that lexical concepts are atomic in that they lack constituent structure; Lila Gleitman's work on syntactic bootstrapping, which highlights the systematic relations between the form and meaning of lexical items; Alan Leslie's focus on the mechanisms underlying conceptual thought; and Susan Carey's work on the acquisition of integer concepts, which emphasizes the importance of distinguishing between representations that have numerical content from number concepts—the idea that form, not just content, matters for theories of conceptual representation. The fragment of the theory of conceptual form developed in this book is in large part an attempt to develop a theory of conceptual representation that incorporates all those ideas.

In writing the book, I have benefited from comments on drafts of chapters from Elizabeth Camp, Susan Carey, Theo Fenton, Dan Harris, Catrinel Haught (-Tromp), Ellen Lau, Eric Margolis, Peter Mazalik, Alexander Noyes, Paul Pietroski, David Poeppel, Lance Rips, Katherine Ritchie, and Tomasz Zyglewicz. I am grateful for their insightful feedback. The book is much improved by my attempt to address their comments. They, of course, cannot be held responsible for any errors that remain. I am particularly grateful to Ellen Lau, Kate Ritchie, and Susan Carey for the detailed feedback that they provided. Ellen read and commented on an early draft of all but the last chapter and used a draft of the first five chapters in a graduate course and shared the feedback that she received. Kate provided me with detailed comments on multiple chapters and invited me to join sessions of a seminar that she was teaching, in which they discussed drafts of a couple of chapters. Susan has encouraged this line of work while providing constructive criticism for the last thirty years, but she also made the time to meet with me to discuss her detailed comments on each of the first five chapters. I greatly appreciate those discussions and helpful suggestions. One suggestion that I did not fully implement, however, was to describe in greater detail the rich empirical basis for many of the claims in the book. I have chosen to refer readers that seek that level of detail to the relevant literature. I thought that a richer presentation of that material would get in the way of a clear presentation of the theory as a whole. I hope that was the right decision.

Bits and pieces of the ideas and arguments that found their way into the book were developed over the years through discussions with numerous colleagues, including David Barner, Paul Bloom, Susan Carey, Greg Carlson, Martin Chodorow, Andrei Cimpian, Andre Eliatamby, Roman Feiman, Rochel Gelman, Susan Gelman, Tamsin German, Lila Gleitman, Geoff Hall, James Hampton, Dan Harris, Paul Haward, Sangeet Khemlani, Melissa Kibbe, John Kim, Joshua Knobe, Ellen Lau, Alan Leslie, Sarah-Jane Leslie, Tania Lombrozo, Eric Mandelbaum, Greg Murphy, Alexander Noyes, Jeff Pelletier, Paul Pietroski, David Poeppel, Jake Quilty-Dunn, Hugh Rabagliati, Marjorie Rhodes, Katherine Ritchie, Steven Sloman, Jesse Snedeker, Michael Strevens, Virginia Valian, Alexis Wellwood, Fei Xu, and Tomasz Zyglewicz. John Kim and David Poeppel get special thanks for listening to my often less-than-half-baked thoughts on various aspects of the theory while also providing useful criticism and encouragement. Numerous conversations with my friend, the poet Michael Scharf, helped me get better at talking about the ideas in this book to an audience of intelligent nonexperts.

And, of course, I am grateful to my many collaborators over the years who have contributed relevant ideas and helped collect data that bear on them, and also to Martin Chodorow, who has provided helpful advice on statistics on many occasions. Feedback from audiences at numerous colloquia and conference presentations, especially at the Society for Philosophy and Psychology, has also played an important role in developing the ideas contained in the book. A workshop on "The Spectrum of Kinds," organized by Carla Umbach and Cecile Meier at the 2024 DGfS (German Linguistic Society), and a workshop titled "Beyond Associations," organized by David Poeppel and Nina Kazanina for The Foundations Institute and the Ernst Struengmann Institute, provided particularly enriching discussions and feedback on key ideas contained in the book.

Virginia Valian gave detailed and very helpful feedback on the book proposal and has helped me think through many of the ideas in the book. I am also grateful to Susan Gelman for putting me in touch with Philip Laughlin at MIT Press with the suggestion that he consider my book proposal. Philip has been super-responsive and great to work with and the reviews of the proposal and the manuscript that he solicited were very helpful for shaping and revising the book.

The psychology department at Hunter College and the linguistics program at the City University of New York (CUNY) have been tremendously supportive of this work and provided me the opportunity to develop these ideas over a long time. I am also grateful to the PSC-CUNY grants that funded most of my research on conceptual form.

This book would not exist without the love, support, and help of my family. My children, Hurley and Iris, provided countless judgments, and my wife, Kendra Hurley, provided encouragement, helped make time for me to write, and gave me detailed suggestions for how to make my writing clearer and more accessible. If the first few chapters read well, that is the result of her efforts. Where they don't, it is despite her best efforts to get me to jettison my native academese.

Given the long gestation of this book, many who I most wanted to share this book with are no longer alive, including John Macnamara, Lila

Gleitman, Jerry Fodor, Sam Glucksberg, and my father, Birendra Prasada. I'd like to dedicate this book to their memory. Noam Chomsky was unable to read the book. Fortunately, I shared an early draft of the first five chapters with him. His response was encouraging: "Seems to me that we are on parallel paths—maybe even better in a non-Euclidean geometry. Very much interested in seeing more."

Here, then, is more of the book than any of you have seen. I hope you find it to be of interest.

I Introduction

1 Situating the Study of Conceptual Form

As you read this book, you will learn about the types of structures that are found in commonsense concepts. If you think about it, that is remarkable! How can I construct complex thoughts about the structure of concepts and convey them to you via language? I cannot answer that immensely interesting and difficult question. However, I do hope to illuminate a key component of the mechanisms involved in accomplishing that feat. Specifically, I intend to illuminate the nature and structure of the *conceptual elements* out of which linguistically expressible complex thoughts are made, as well as how the structure of those conceptual elements guides how we think, reason, and talk. As we will see, even the simplest conceptual elements such as the concepts DOG, BOTTLE, and TREE possess intricate and abstract structure. This intricate and abstract structure—what I call *conceptual form*—is generally hidden from us. It is not perceived, and we are not conscious of it in normal circumstances. Conceptual forms function like mental lenses that specify the perspectives that a concept provides for thinking, talking, and reasoning about things and their properties. The concept DOG, for example, provides a perspective or "mental lens" that allows us to use it to think of Fido not simply as a single entity, but as *one of an unlimited number of things of the same kind* (i.e., dogs); and to explain some of his properties by reference to the kind of thing he is (e.g., Fido has four legs *because* he is a dog); and to have normative expectations for those properties (e.g., Dogs are *supposed to* have four legs, and there is something wrong with a dog that doesn't have four legs). The concept DOG also allows us to think thoughts like DOGS ARE WIDESPREAD, which is a thought about the kind, or the category as a whole.[1] Individual dogs cannot be widespread. These are ways of thinking and talking that the mental lens provided by concepts like DOG make available. They are not properties of the things that the concept represents (dogs). Concepts direct our attention to the things that they represent, so we are aware of the perspectives that the concept provides only incidentally—in the implications that the perspectives have for how we think and talk about the things that the concept represents.

A central goal of this book is to present a theory of the mental lenses that are naturally available to us for thinking and talking—a theory of conceptual form. My focus is on the structure of *lexically expressible concepts* because these are the concepts that can be combined in systematic ways to yield indefinitely many arbitrarily complex, precise, and novel thoughts about an unlimited range of topics that can be expressed via language. That is a feat that every normally developing child is capable of, but no animal is. As such, the theory of conceptual forms reveals an important component of distinctively human thought.

Existing approaches to conceptual representation have generally not investigated the perspectives or mental lenses that concepts provide. As a consequence, existing approaches to conceptual representation fail to provide an account of many phenomena that depend on the perspectives provided by concepts. A sample of the types of phenomena that are explained via the conceptual form of concepts, but are unexplained by existing approaches, is given in table 1.1. Although their dependence on conceptual form may not be immediately apparent, it will become evident in the course of this book. It is important to note that although each phenomenon listed in table 1.1 is illustrated via a single concept, all the phenomena hold for systematic but nonobvious *classes* of concepts that share the relevant conceptual form. There are many more phenomena that depend on the conceptual form of concepts, but the sample in table 1.1 should provide a sense of the diversity and depth of the phenomena that depend on the perspectives provided by the concepts that we use to think and talk.

The phenomena in table 1.1, and many others like them, come into focus only when we acknowledge that concepts not only represent information about category members, but also provide perspectives from which to think and talk about things and their properties. In this discussion, I briefly explain why existing work on conceptual representation does not shed

Table 1.1
A small sample of phenomena that depend on the conceptual form of a concept

Conceptual Form Is Needed to Explain:
(1) How we can think of Fido as *one* of an *unlimited* number of things of the same kind (called "dogs")
(2) How we can use the concept DOG to think about specific members of a kind such as Fido, Spot, and Jett, as well as thoughts like "Dogs evolved from wolves," which are about the kind itself
(3) Why we can easily understand a fairy tale in which a pumpkin is turned into a carriage, but not one in which a pumpkin is turned into a different pumpkin
(4) How we can characterize kinds, and, how we can have multiple ways of characterizing these never-encountered abstract entities that have an unlimited number of instances in them
(5) Why there are different constraints on the kinds and properties that can be expressed via different linguistic expressions for characterizing kinds
(6) Why, although it may be odd, the question "Is squirrel healthier than raccoon?" makes sense, but the question "Is chair denser than table?" does not
(7) Why "Lions have manes" strikes us as true, while "Lions are male" strikes us as not merely false, but nonsensical
(8) Why we can use "newspaper" and "magazine" to refer to the organizations that publish the respective reading materials, but not "book" to refer to the author of a book
(9) Why we think that something cannot simply be an animal without being a specific kind of animal
(10) Why metaphorical thought is possible
(11) Why "Men are dogs" is a metaphor, but "Cats are dogs" is gibberish
(12) Why "Some cats are dogs" is a metaphorical statement, but "Cats are dogs" is gibberish.

Note: Each phenomenon holds for systematic but nonobvious classes of concepts that share the relevant conceptual form.

light on the perspectives provided by commonsense concepts. My consideration of existing approaches to conceptual representation is limited to a few key ways in which they contrast with the approach taken in this book to help situate the theory of conceptual form. Readers interested in more comprehensive discussions of approaches to conceptual representations may consult Medin (1989), Hampton (1997), Murphy (2002), Goldstone, Kersten, and Carvalho (2018), or Margolis and Laurence (1999, 2015, 2023) among others.

1.1 Standard Approaches to Conceptual Representation Are Nonperspectival

The vast majority of empirical work on conceptual representations has focused on concepts of ordinary objects like tables and dogs. Despite several decades of research, there is no consensus on how even such concepts are represented. Nevertheless, psychologists generally agree that concepts are mental representations; categorization is a central, if not the primary. function of object concepts like TABLE and DOG; and categories are represented via summary descriptions. Approaches differ, however, with respect to the type of summary description they posit for representing categories.

The classical theory of concepts proposes that categories are represented via summary descriptions that are definitions (Margolis & Laurence, 1999). Definitions are understood to consist of a list of properties that are individually necessary and jointly sufficient for membership in the category represented by the definition. Although intuitively appealing, it quickly became evident that it is not possible to have definitions for all but a few categories. You can get a sense of why this is by considering which properties may be included in the definition of dogs. Virtually any property that one may consider for inclusion in the definition, such as BARKS, HAS TAIL, HAS FOUR LEGS, and IS FRIENDLY, turns out not to be a logically necessarily property of dogs and thus cannot be included in the definition. There is no contradiction in thinking of a dog that does not bark, does not have a tail, does not have four legs, or is not friendly. In fact, there are actual dogs that lack each of these properties. On the other hand, properties that are necessary, such as BREATHES, GROWS, and COMES INTO EXISTENCE VIA REPRODUCTION, are shared by many categories and cannot, individually or in combination, serve to distinguish dogs from nondogs. Similar difficulties are encountered when trying to formulate a definition of virtually any category other than mathematical concepts, and thus it is now generally assumed that concepts are not represented via definitions.

The prototype approach proposes that categories are represented via summary descriptions called *prototypes* (Rosch, 1975, 1978). Prototypes consist of a list of the properties that are typical or characteristic of members of a category. Properties that are part of the prototype can be properties that are true of all members of a category; however, crucially, they need not be. Returning to the concept DOG, properties like BARKS, HAS TAIL, HAS FOUR LEGS, and IS FRIENDLY can all be included in the prototype, as they are typical of members of the category even though they are not necessary for belonging in the category. The typicality of a property, and thus the likelihood of it being included in the prototype that represents the category, is influenced by a variety of factors, including how prevalent a property is in members of the category, how distinctive it is of the category, and the extent to which we think that it is ideal for members of a category (Barsalou, 1985; Murphy, 2002; Burnett, Medin, Ross, & Blok, 2005).

Starting in the 1980s, researchers began to highlight the explanatory dimension of concepts. Concepts, in this view, are not merely devices for sorting things; they are centrally involved in understanding and explaining why things have the properties they do, how those things and properties come about, and why things behave and change in the ways that they do and not other ways (e.g., Carey, 1985; Murphy & Medin, 1985; Keil, 1989). The explanatory dimension of concepts has been incorporated into theories of conceptual representation in two distinct ways—one involving summary representations and another that eschews summary representations. I discuss the latter approach in the next section. With the former approach, categories are represented via a summary description of the causal-explanatory structure that is shared by category members (e.g., Ahn, 1999; Rehder, 2003, 2017; Sloman, Love, & Ahn, 1998). Those summary descriptions contain not only category-typical properties but also the causes that explain those properties, relations between properties, and relations between relations (Medin, 1989). As such, the summary descriptions with this approach are not merely descriptions of the properties shared by category members. They also describe the causal explanatory structure that is shared by category members.

Finally, the exemplar approach denies the idea that categories are represented via summary descriptions. It proposes instead that categories are represented via memory representations of individual examples of the category (Medin & Schaffer, 1978; Nosofsky, 1988).[2] So, for example, the category "dog" is not represented by a summary mental representation DOG, which contains properties shared by members of the category, but instead is represented by mental representations of a number of exemplars of the category (e.g., Fido, Spot, Rover, Lassie) and each of their properties.

Although there are important differences between these approaches, as well as the many versions of each of these approaches, they all share the characteristic that they are focused on representing properties of category members, which are implicitly assumed to be transparently represented by the concept. The exemplar approach represents the properties of various category members individually. The other approaches represent a summary of the properties that are generally shared by members of the category.

The approaches differ in whether they focus on the statistical, causal, or logical structure of the things that they represent. However, in each case, the structure that the concept represents is simply the structure of the members of the category.

None of these approaches investigate whether the form of the conceptual representation itself functions as a mental lens that contributes perspectives for thinking and talking about things and their properties. In other words, the formal structure of the concept (the structure of the representation itself) has no place in these approaches. Consequently, the conceptual representation cannot provide perspectives for thinking and talking about the things that it represents.

Except for the theory view of concepts, the approaches to conceptual representation described here were primarily developed in the context of categorization studies with adult humans. Over the past several decades, concepts have been investigated from many more directions, including comparative, neuropsychological, computational, linguistic, and philosophical approaches (Margolis & Laurence, 2015). This has led to many new findings, insights, and theoretical questions and advances. However, the new approaches have *not* yielded new ways of representing a category. They generally implicitly or explicitly assume one or more of the standard approaches to conceptual representation described here.[3]

Embodied or grounded cognition approaches to cognition, for example, have questioned the idea that cognition simply involves amodal, abstract conceptual representations (e.g., Barsalou, 2008, 2012; Kiefer & Pulvermuller, 2012). In such views, the concept DOG, for example, is not simply an amodal abstract characterization of dogs but also includes modal sensorimotor representations of what dogs smell, feel, and look like; our affective reactions to them; and the motor movements involved in interacting with dogs, such as petting them. All this modal and amodal knowledge is proposed to be activated and used when thinking about dogs. As such, embodied approaches greatly expand the types of information that are encoded in a concept.

But crucially, for our purposes, these approaches, like previous approaches, do not seek to characterize the perspectives that concepts provide. They merely expand the content of those concepts to include modal sensorimotor and affective representations that enter into the experience of category instances. Typically, the modal and amodal features activated by a concept are understood to be either properties of a particular exemplar or as typical of members of the category. As such, research within the embodied approach to conceptual representation tends to assume that concepts have the structure posited by the exemplar- and prototype-based views of concepts. Whatever merits the embodied approach to concepts may have, it does not seek or provide an account of the mental lenses that concepts provide.

In fact, as we will see, the investigation of conceptual form shows that even concrete object kind concepts like DOG have intricate, abstract, and formal structures that are abstract along multiple dimensions, suggesting that the abstract, amodal representations involved in concrete concepts like DOG are much richer than previously thought.

Computational models of conceptual representation have also become increasingly prevalent over the past few decades. For the most part, these models investigate sophisticated variants of the similarity-based prototype or exemplar approaches to conceptual representation. They have revealed interesting consequences of different types of feature representations, learning rules, and ways of computing similarity and extracting similarity and statistical structure within and across categories. However, like their noncomputational counterparts, the focus is on discovering the structure shared by the instances in a category that supports categorization, and they do not seek to model the perspectives provided by different classes of concepts.

Finally, the Bayesian program learning (BPL) framework provides an exciting new framework for investigating theories of conceptual representation (Lake, Salakhutdinov, & Tenenbaum, 2015). The framework views concepts as probabilistic, generative programs and has been used to learn visual concepts. Thus far, it has not been used to acquire or represent the structures that encode the perspectives provided by lexically expressible kind concepts, but because it represents concepts as structured procedures, it holds the promise of doing so. In chapter 11, I provide a more detailed description of this framework and how it may be expanded to represent and acquire the perspectives provided by lexically expressible concepts.

1.2 A Notable Exception: Framework Theories

The theory approach to concepts holds that our commonsense concepts are not isolated mental representations constituted of a small number of features (primitive concepts); instead, they form interconnected mental structures that function as intuitive theories that allow us to understand and explain the world around us. It is also generally assumed that the intuitive theories and conceptual knowledge that we acquire are constrained by domain specific framework theories. Framework theories identify domains in three interrelated ways: via (1) the type of phenomena to be explained, which (2) involve a domain of ontologically distinct entities, and (3) the type of explanatory mechanism or principle required for providing coherent explanations for the phenomena (Wellman & Gelman, 1992).

Although early work within the theory approach focused on causal mechanisms (Carey, 1985), there is now a recognition that our conceptual systems use multiple modes of construal (Keil, 1994) that support not only causal explanations, but also teleological and formal explanations (e.g., Springer & Keil, 1991; Kelemen, 1999; Lombrozo & Carey, 2006; Prasada & Dillingham, 2006; Rivera, Prasad, & Prasada, 2023).

Framework theories do not specify particular concepts like DOG, TABLE, or TREE. Instead, they provide frameworks for developing specific intuitive theories involving specific concepts that can be used to think about specific kinds of things within the domain. Framework theories are "both open-ended and constraining; they allow and inspire the development of more specific theories" (Wellman & Gelman, 1992, 342). Framework theories thus are crucial to the acquisition of specific intuitive theories and specific concepts. As such, framework theories may be thought of functioning as mental lenses through which intuitive theories and some kinds of conceptual knowledge are acquired.

Furthermore, because different modes of construal can be used to think about the same entities, framework theories that exploit different modes of construal may be thought of as providing different perspectives for thinking about the same thing. As such, framework theories, unlike other approaches to conceptual representation, are perspectival.

Although framework theories are perspectival, many questions about framework theories currently lack answers. Most important, it is unclear how framework theories are mentally represented. What is the form of the mental representations that constitute a framework theory? How are those representations accessed and used in the formation of specific intuitive theories? What are the formal characteristics of the concepts that are acquired via a framework theory?

Further articulation of our understanding of framework theories will help address many of those questions. Nevertheless, framework theories are not well suited for explaining the types of perspectival phenomena listed in table 1.1. There are two reasons for that. First, framework theories have the wrong *granularity* for explaining such phenomena. As we will see in sections 1.5 and 7.7, the class of kind concepts that contains concepts such as NEWSPAPER, DICTIONARY, and ENCYCLOPEDIA provides a different way of thinking and talking about instances of kinds than the class that contains concepts such as MAGAZINE, BOOK, and PAMPHLET. Framework theories do not seem to be suited for developing such fine-grained perspectival differences. Second, and more important, framework theories *presuppose the perspectives provided by kind representations*. A given framework theory guides the acquisition of intuitive theories and concepts for certain kinds of things/stuff. For example, we have different framework theories for developing understandings of kinds of physical objects, kinds of living things, and human beings (Wellman & Gelman, 1992). Whereas a framework theory guides the acquisition of conceptual knowledge for the kinds of things/stuff in a given domain, kind representations are needed for representing kinds of things/stuff in all domains. As such, the perspectives provided by kind representations cannot be provided by framework theories but are instead presupposed by framework theories.

We can get a better sense of how framework theories can provide perspectives, as well as the limits on the perspectives that framework theories can provide, by considering the framework theory that has been studied in the greatest detail—psychological essentialism.

1.3 The Perspective Provided by Psychological Essentialism

Research on psychological essentialism has investigated the hypothesis that the conceptual representations of natural kind categories and some social categories provide an essentialist perspective for thinking and talking about category members and their properties. According to psychological essentialism, in thinking and talking about certain kinds of things, we assume that their observable category-typical characteristics are caused by an underlying essence which, among other things, determines the thing's true identity. Although we typically do not know the essence of any given category, the conceptual representation of the category has a structure within which there is a placeholder for the essence (Gelman, 2003, 2004; Medin & Ortony, 1989). This conceptual structure underpins a perspective in which we have the intuition that there is an essence to be discovered; think of members of the category as remaining the same kind of thing even as many of their properties change (e.g., Fido can change in size, color of coat, and friendliness and remain the same dog); assume sharp boundaries for categories (e.g., something either is or is not a dog—it can't be partly a dog); expect to discover nonobvious properties that are shared by category members; expect shared properties to generally be biologically determined rather than due to nurture; and readily extend properties that we think are caused by the essence to all category members (Gelman, 2003). This essentialist perspective is found in the ways in which adults and children think and talk about animals, plants, substances, and gender categories, but not artifacts (Gelman, 2003, 2004). Furthermore, the essentialist perspective is found even in cases in which the explicitly held views of a culture conflict with that perspective (Astuti, Solomon, & Carey, 2004).

The essentialist perspective is understood to be a consequence of the structure of the conceptual representation itself (Gelman, 2004). It is not a property of the members of those categories. Examining members of the category would not reveal that they share an essence. It is a way of thinking and talking about the members of those categories and their properties that is a consequence of the structure of the conceptual representations to think and talk about certain categories.[4] That this perspective is a consequence of the structure of our concepts rather than a property of the members of the categories is highlighted by the fact that the essentialist perspective functions as a barrier to a scientific understanding of biological categories and their evolution (Hull, 1964; Gelman & Rhodes, 2012).

1.4 Beyond the Essentialist Perspective

As important as the essentialist perspective is to commonsense conception, it is but one perspective of the myriad perspectives that are provided by our commonsense concepts. The theory of conceptual form seeks to account for the full range of perspectives provided by our commonsense concepts, including the essentialist perspective.

The essentialist perspective itself arises from two assumptions—an essence assumption and a kind assumption—that are embodied in the structure of the conceptual representations of some categories (Gelman, 2004). Research on psychological essentialism has presupposed kind representations and investigated the perspective that results from the essence assumption. As such, a full theory of the essentialist perspective awaits a theory that specifies how the kind assumption is embodied in the structure of concepts and makes available the perspectives that allow us to think and talk about things as instances of a kind, as well as about the kinds themselves. I develop exactly such a theory in chapter 4 and show how that theory generates a large variety of perspectives that are provided by lexically expressible kind concepts in chapter 5.

As we will see, all kind concepts provide perspectives for thinking and talking about kinds and their instances, but each formally distinguished class of kind concepts provides a distinct way in which those perspectives are realized. Each of the concepts DOG, NECKLACE, BOOK, PAPER CLIP, POEM, OBITUARY, ANIMAL, ROOK, and WOOD, for example, belongs to a distinct class of kind concepts, with each of those classes providing a distinct perspective for thinking about the kinds and instances in that class.

Although the different classes of kind concepts provide a broad range of perspectives, the perspectives that they provide do not vary in arbitrary ways. Rather, they differ in systematic ways along a few key dimensions. One dimension on which kind concepts differ is in whether the perspective provided by the kind specifies what counts as one instance or not and whether it provides more than one way of specifying what counts as one instance. Concepts such as WOOD or SAND, for example, do not determine what counts as an instance of that kind. To identify an instance of those kinds, we need the support of another concept, such as PIECE, PORTION, or PILE. Furthermore, what counts as a piece, portion, or pile is not determined by being wood or sand. Consequently, we can choose to identify what counts as one instance of wood or sand differently on different occasions depending on our interests. All of this contrasts with the perspective provided by concepts such as TABLE or DOG that *do* determine what counts as one instance. We do not need the support of another concept to determine what counts as one instance of these kinds. If something is an instance of the kind TABLE or DOG, it is *one* instance of that kind, and different contexts do not allow us to count instances differently. Kind concepts like BOOK and MAGAZINE differ from both of those types of kind concepts because they provide a perspective in which the kind determines what counts as one instance in *two* ways. For example, if I buy five copies of this book and someone wants to know how many books I bought, the answer is, obviously, "five." If, on the other hand, someone wants to know how many books I have written, the answer is one, not five (or however many copies have been published), as this is the only book that I have written. These observations show that one dimension on which classes of kind concepts differ is in whether the perspective provided by the kind concept individuates its instances (i.e., specifies what counts as one instance) and the number of ways in which the kind concepts in that class individuate their instances.

It is worth noting that existing approaches to conceptual representation focus only on what is *shared* by instances of a category, and, as a consequence, they do not have a way of representing how a kind concept individuates its instances, as that requires specifying the ways in which the kind's instances intrinsically *differ* from one another (more on this in chapters 4 and 5). Other dimensions along which the perspectives provided by different classes of kind concepts vary are discussed in chapter 5. There, I develop a fragment of the generative system that yields the constrained variety of perspectives provided by distinct classes of lexically expressible kind concepts.

1.5 Investigating Conceptual Form

Although psychologists tend to think that the meanings of words are represented by concepts, they have generally assumed that conceptual representations must be characterized independent of language. This methodological stricture has prevented investigation of the perspectives provided by lexically expressible concepts. Some of the clearest evidence concerning the perspectives that concepts provide comes from the ways in which the perspectives are reflected in language. Here, briefly, are some ways in which the perspectives provided by kind concepts can be investigated empirically.

As already mentioned, under normal circumstances we are not aware of the conceptual form of a concept or the perspectives that it provides. One way to make characteristics of conceptual form evident is to hold the object of conception constant and observe how the way we think of it and its properties change when we use different concepts to think about it. For example, we can think of a given thing as a table, or as a hunk of wood, and note how the change in perspective affects the ways in which we think and talk about it. The most obvious difference in this case is in how we think about the thing's structure or shape. In thinking of the thing as a table, we think of its structure/shape as being nonarbitrary—it has the structure that it does *because* it is a table, and it would not be a table with just any structure. On the other hand, when we think of the thing as wood, we think of its structure as being arbitrary—it does *not* have the structure that it does because it is wood, and it would still be wood if its structure were completely different. The structure of the entity is, of course, simply what it is—it is neither arbitrary nor nonarbitrary. It is the concept TABLE that represents its structure as being nonarbitrary, and it the concept WOOD that represents its structure as being arbitrary.

Other differences in the perspectives provided by the two concepts include the fact that the concept TABLE represents the thing as one instance of the kind, whereas WOOD does not, on its own, represent it as one instance, but instead represents it as a continuous amount of the kind; TABLE, but not WOOD assigns it an orientation, so it makes sense to ask whether the table is upside down, but not if the wood is upside down; TABLE, but not WOOD assigns it a part structure, and thus it makes sense to ask how many parts the table has but not how many parts the wood has; and we can use the quantifier "many," but not "much," to talk about quantities of tables but the opposite holds when talking about quantities of wood. Because the object of conception is held constant, these differences are understood as being due to the perspective that each concept provides for thinking about that thing. Linguistic evidence for the proposed perspectives can be complemented by experiments that test predictions concerning the context in which one or another perspective will be chosen (e.g., Prasada, Ferenz, & Haskell, 2002; Li, Dunham, & Carey, 2009).

Our example illustrates a key element of investigating conceptual form, which is to identify *concept-dependent features* such as the proper orientation, mode of quantification, or part structure. These features cannot be understood as characterizing an entity independent of our conceiving of the entity via a concept of a given type. Not all features are concept dependent. For example, the color and weight of the thing that is conceived of as either a table or some wood does not depend on whether it is conceived of as a table or as some wood. Concept-dependent properties, obviously, cannot serve as the basis of categorization. They are a consequence of categorization and part of the perspective that the concept provides for thinking about the thing.

As discussed, traditional approaches to conceptual representation are generally focused on specifying the properties by which things are identified as belonging to a given category. As such, that research has generally not considered whether properties are concept dependent. However, identifying concept-dependent properties is a key element of specifying the conceptual form of a concept.

Minimal pairs of lexical items that have intuitively similar meanings and behave similarly in many contexts, but support very different interpretations in other sentential contexts, provide an opportunity to investigate nonobvious meaning and perspectival differences between lexically expressible concepts. Consider the words "magazine" and "newspaper," which have intuitively similar meanings and generally appear in similar sentential contexts but nevertheless support very different interpretations in the following sentences:

1. John will read the newspaper before cooking dinner, as will Bill.
2. John will read the magazine before cooking dinner, as will Bill.

In sentence 1, John and Bill may read the same newspaper (e.g., *The New York Times*) and it may even be the same copy of that newspaper, but it need not be the same newspaper. Alternatively, John may read *The New York Times* and Bill may read *The New York Daily News*. In contrast, in sentence 2, John and Bill must read the same (previously identified) magazine, such as the May issue of *Spin*, or even the same copy of the magazine. It cannot be that John reads *Spin* and Bill reads *Vogue*. These linguistic facts suggest that there is some nonobvious difference in the way that we think and talk about magazines and newspapers that gives rise to this interpretive difference. That this is not a mere idiosyncratic difference between "newspaper" and "magazine" is suggested by the fact that in this type of context, "book," "pamphlet," and "journal" are all understood in the same manner as "magazine," whereas "dictionary," "encyclopedia," and "map" are understood in the same manner as "newspaper."

In chapter 7, I propose how and why the perspectives provided by these two classes of concepts differ, as well as how that explains the difference noted here. For now, I simply want to use this example to illustrate how nonobvious differences in the meanings and perspectives provided by different classes of kind concepts can come to light when we consider the ways in which they can contribute to sentence meaning. So long as conceptual representations are investigated independent of language, these aspects of lexically expressible concepts are likely to remain undiscovered.

In addition to directing attention to nonobvious aspects of a concept's structure, linguistic evidence can help reveal formally distinct ways in which concepts can be connected to one another. For example, even though we can use a bare plural generic to characterize tires as being round (3) or as being black (4), we can use an indefinite singular generic to characterize tires as being round (3′), but not as being black (4′):

3. Tires are round.
4. Tires are black.
3′. A tire is round.
4′. A tire is black.

This difference suggests that we represent the connection between being a tire and being round in a different way than the connection between being a tire and being black. Similar patterns are found if we consider the kind dog and the properties of having four legs and of wearing a collar, the kind architect and the properties of designing buildings and writing neatly, or many other kinds and properties across content domains. Because similar contrasts are found across content domains and properties, it suggests that the two ways of connecting properties must be characterized in content-independent, formal terms. Prasada and Dillingham (2006, 2009) show that two types of formal connections—principled and statistical connections—underlie the linguistic phenomena in (3)–(4), as well as a number of other linguistic and nonlinguistic differences (see chapter 6).

These are just some examples of how systematic differences in the ways in which we talk about things can be used to investigate aspects of conceptual structure that are not immediately apparent or intuitively obvious. Linguistic evidence, like any type of evidence, doesn't directly identify the conceptual structures underlying the linguistic phenomena; instead, it directs us to try to formulate hypotheses concerning the nature of the formal structure that would explain the observed difference. Those hypotheses can then be tested via experiments or by looking at their predictions about further linguistic judgments. That is a strategy that I exploit throughout this book.

Finally, in developing a fragment of the formal structure of kind representations in chapters 4 and 5, I begin with some core intuitions that appear to characterize the ways in which we think and talk about kinds and their instances. Those intuitions inform the formal structure hypothesized. In many cases, those intuitions have been corroborated by experimental data. Other cases await corroboration. The discovery and development of the theory of conceptual form constitute an ongoing and thoroughly empirical enterprise.

In seeking a theory of the formal structure of concepts and using linguistic phenomena as one source of evidence regarding conceptual form, I am following in the footsteps of Ray Jackendoff (1983, 1987, 2002). His work on conceptual structure has been an important inspiration and influence on the work in this book. Although the theory of conceptual form shares the goal of specifying the formal structure of commonsense concepts with Jackendoff's theory of conceptual structure, there are fundamental substantive differences in the formal structure of kind and instance-of-kind representations developed in this book and the formal structure of type and token representations in the work of Jackendoff (and many others). I outline those fundamental differences in chapter 4 after I have specified the formal structure of kind concepts.

1.6 Cross-Linguistic Variation and Whorfian Worries

It is well known that languages differ in which meanings they lexicalize and how they lexicalize meanings. English has a different word for a brother and a (male) cousin, but Hindi uses the same word, *bhai*, to refer to both brothers and male cousins. And whereas "furniture" is a mass noun in English, *meuble*, the corresponding word in French, is a count noun. And, of course, the syntax of languages also differ. Given all this, one might wonder whether the decision to use language to investigate conceptual form commits us to the idea that speakers of different languages may think in fundamentally different ways.

There are several reasons why that conclusion does not follow. First, when linguists Edward Sapir and Benjamin Whorf first made their claims that language shapes thought, it was believed that languages were very different from one another and could differ from one another in arbitrary ways. Given that belief, it was plausible to think that speakers of different languages may think in fundamentally different ways. However, linguistic research over the past seven decades has shown that, in fact, languages are deeply similar in their abstract formal structure and much of that structure is not evident in the surface form of expressions (Adger, 2019; Chomsky, 1986, 2023). Given the finding that all languages are deeply similar in their nonsurface properties, the potential for language differences leading to significant differences in how we think is no longer plausible.

Second, many discussions of language differences overestimate how different languages are and also underestimate the diversity found within a single language. Take the observation that "furniture" is a mass noun in English but the word for it in French, *meuble,* is a count noun. While it is worth noting this difference and investigating whether there are any differences in the perspectives that the two words provide for thinking and talking about furniture, it should also be noted that such variation also exists *within* a language. "Clothes," "clothing," and "garments" are used to talk about the same things in English. Although there may be subtle differences in the perspectives that each of these words provides for thinking and talking, they are surely minor and not evident in a wide range of contexts. We can expect the same when we consider such differences across languages rather than within a single language.

Third, the absence of a transparent way of marking a distinction within a language does not mean the conceptual distinction is not available and easily accessed in relevant contexts. Recall the example of "magazine" and "newspaper" discussed earlier in this chapter. Despite the lack of a surface way of marking the different perspectives that they provide for thinking and talking, those differences are readily available in contexts such as (1) and (2). What is true within a language also holds across languages. So even though Hindi speakers have no lexical way of distinguishing between brothers and male cousins, the conceptual distinction is readily available to them and used in relevant contexts. Experimental investigations of the conceptual representations of object kinds and substance kinds in languages such as Mandarin and Japanese, which do not mark the count-mass noun distinction and thus do not talk about quantities of objects and quantities of substances in systematically different ways, and languages like English, which do mark this distinction, tell a similar story. Children and adults using both types of languages understand instances of object kinds to have nonarbitrary structure and instances of substance kinds to have arbitrary structure, and use the same types of cues to the nonarbitrariness of structure to learn names for object kinds and substance kinds (Imai & Gentner, 1997; Li et al., 2009; Prasada et al., 2002). What differences are found between language types in this research are best explained as effects of language on language rather than effects of language on thought (Li et al., 2009).

Finally, while cross-linguistic differences in lexicalization and grammatical structure are well attested, we do not yet know about the extent to which the range of perspectives made available by classes of lexically expressible concepts vary across languages. There are good reasons to think that such perspectives will be largely shared across languages, but that is a question for future research. What is clear, though, is that using language as a tool to investigate conceptual form does not commit us to linguistic relativity in any important way.

2 The Perspectival Nature of Commonsense Concepts

Lexically expressible concepts—such as DOG—function as mental lenses that make available specific perspectives for thinking about things and their properties. How is that possible? How is it that a lexically expressible concept contains information about things and their properties, but also specifies a perspective that tells us how to think about things and their properties? Is the perspective provided by an individual concept, such as DOG, specific to that concept, or is it shared by a whole class of concepts? And do our conceptual systems provide just a few general perspectives shared by large numbers of concepts, or do they make available an abundance of perspectives? Finally, how do the perspectives provided by lexically expressible concepts differ from the perspectives provided by complex concepts that are expressed via phrases or sentences? I address such questions about the general character of the perspectives provided by lexically expressible concepts in this chapter. In chapters 4 and 5, I will develop a proposal that specifies the perspectives provided by a wide variety of classes of lexically expressible kind concepts.

2.1 The Perspectives Provided by Commonsense Concepts Are Abstract

Even concepts such as DOG and BOTTLE, which provide the means for thinking about the concrete objects that we encounter in life, provide multiple abstract perspectives from which to think and talk. For example, two distinct but intrinsically connected perspectives that are provided by the concept DOG are the instance-of-kind perspective and the kind perspective. These two perspectives license different ways of thinking and talking. Nevertheless, the two perspectives are intrinsically connected. You cannot have one without the other—kinds are understood to have an unlimited number of instances of that kind in them, and instances are instances of a given kind.

The kind perspective provided by DOG allows us to think and talk about an abstract entity—the kind dog, as such. The kind dog, unlike its instances, cannot be perceived, although we can think and talk about it. The thought that dogs evolved from wolves involves a thought about how these two kinds of things are related, not how the instances of these kinds are related to one another—predicates like *extinct* and *invented* are licensed by the kind perspective.

The concept DOG also provides an instance-of-kind perspective, which allows us to think and talk about a particular thing as one of an unlimited number of things of a specific kind. Although I may form the concept DOG on the basis of an encounter that I have with a particular dog (e.g., my dog Jett), the instance-of-kind perspective allows me to think and talk about an unlimited number of other particular things as dogs—both actually existing dogs and potentially existing dogs. For example, the concept DOG (along with other concepts) allows me to wonder whether the next dog that is born will be the same color as Jett.

The instance-of-kind perspective provided by DOG also leads us to think of some of the properties that an instance of the kind (e.g., Jett) possesses as being true of it by virtue of its being the kind of thing it is. For example, we think that the four-leggedness of Jett is a characteristic that he has by virtue of his being the kind of thing he is—namely, a dog. This is why if someone were to ask, "Why does Jett have four legs?" answering, "Because he is a dog" is perfectly natural. As such, concepts such as DOG provide a perspective in which there is an intrinsic connection between categorization and explanation—thinking of something as an instance of a kind via a concept such as DOG not only categorizes the thing but also renders some of its properties intelligible.

The properties of a thing that are rendered intelligible by the kind of thing that something is are also properties for which the instance-of-kind perspective provides normative expectations. We think not only that Jett has four legs because he is a dog, but that he, as well as other dogs, are *supposed to* have four legs, and if they don't, there is something wrong with them. Thus, a dog with three legs is thought not merely to be atypical, but deformed. As such, the perspective provided by DOG intrinsically combines the descriptive and the normative by providing a way of thinking about what dogs are in terms of the properties that dogs are supposed to have.

These observations illustrate the fact that a concept such as DOG provides both an instance-of-kind perspective, which allows us to think and talk about concrete things; and a kind perspective, which allows us to think and talk about an abstract entity—the kind that those concrete objects are understood to belong to. Although distinct, these perspectives are intrinsically connected.

The observations also show some of the ways in which the perspectives provided by concepts such as DOG are abstract. Although we can perceive an instance of a kind, we cannot perceive the thing as one of an unlimited number of instances of a given kind. Similarly, although we can perceive a dog and see that it has four legs, we cannot perceive that it has four legs *because* it is a dog; or that it is supposed to have four legs, and if it doesn't, it has something wrong with it. And, of course, we cannot perceive kinds themselves. All these facts reveal the abstract nature of the kind and instance-of-kind perspective.

Although I have chosen the concept DOG for illustration purposes, both the kind and instance-of-kind perspectives are also provided by CAT, TABLE, TREE, and all other basic-level kind concepts. As such, the perspectives are characterized independent of the concrete properties of any given kind of thing. Furthermore, the perspectives involve abstract entities (e.g., kinds, instances of kinds) and abstract relations (e.g., numerical, explanatory, and normative relations) that are not perceivable (Prasada, 2021b).

2.2 The Perspectives Provided by Commonsense Concepts Are Intricate and Varied

Although all the concepts for basic-level object categories provide the abstract perspectives outlined here, there are systematic differences in the additional perspectives provided by different subclasses of concepts. To help illustrate the variety and intricacy of the perspectives provided by various classes of kind concepts, we will consider the concepts LION, FORK, NECKLACE, and BOOK—each of which belongs to different classes of kind concepts. As we will see, and as summarized in figure 2.1, *the perspectives provided by distinct classes of kind concepts overlap with one another and are distinguished from one another in a variety of ways.*

If we consider two lions or two forks, it is obvious that there is no sense in which we can think of them as the same lion or the same fork. We can think of numerically distinct instances to be the same kind of lion or the same kind of fork, but not the same lion or the same fork. This contrasts with the perspectives provided by NECKLACE and BOOK. We can sensibly think and say that Sally and Sue are wearing the same necklace (when each one is wearing a necklace, not when one giant necklace is around both their necks!); or that they are reading the same book (where each one is reading a different copy of the book). Note that in these cases, we do not simply think that Sally and Sue are wearing the same kind of necklace (e.g., a gold necklace) or reading the same kind of book (e.g., a mystery novel)—we think that they are wearing the same necklace or reading the same book. It appears that whereas LION and FORK provide only a single way of numerically distinguishing lions and forks via the matter that constitutes their instances, NECKLACE and BOOK provide an additional abstract way of distinguishing necklaces and books, such that numerically distinct instances can share that abstract distinguishing character (see figure 2.1). Thus NECKLACE and BOOK provide perspectives in which instances of the kind can be simultaneously thought to be the same and different instances.

However, the concepts NECKLACE and BOOK also differ from one another in the perspectives that they provide. The abstract distinguishing character of necklaces involves perceptual characteristics of the necklaces that it distinguishes. But the abstract distinguishing character of books involves its intelligible character—how it is understood, not its perceptual characteristics. Thus, if Sally and Sue are wearing the same necklace, the necklaces that they are wearing must be perceptually similar; they don't have to be perceptually identical, but they do need to be perceptually similar. On the other hand, if they are reading the same book, the copies of the books they are reading need not be perceptually similar—one may be a paperback, the other hardcover; one may have illustrations, the other not; they may be printed in different fonts using different colors of ink; they could even be in different languages. What makes them the same book is that they contain the same intelligible material.

Returning to LION and FORK, we note that both concepts (and all other concepts for basic-level object categories) provide a perspective from which their instances are understood to differ from one another in accidental rather than principled ways. However, the concept LION, but not the concept

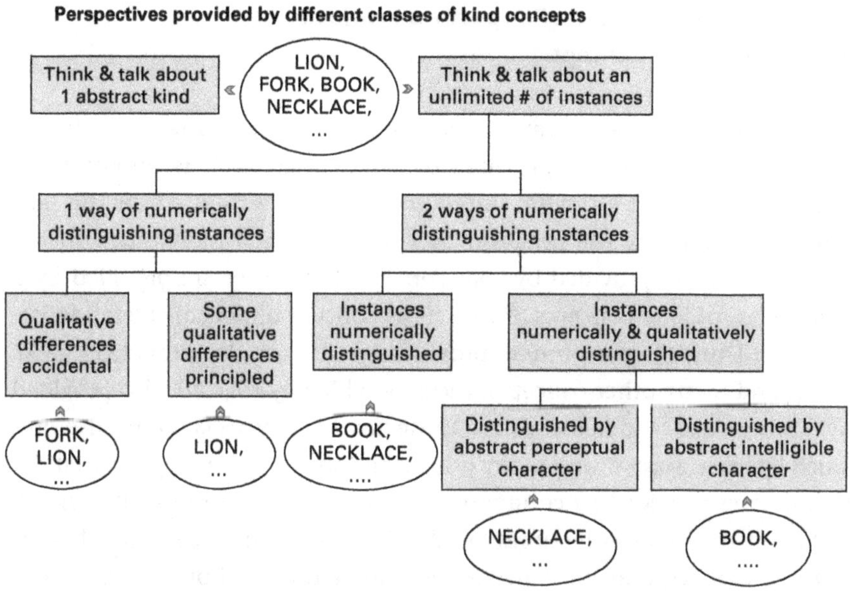

Figure 2.1
Illustration of the overlapping and distinguishing perspectives provided by the various classes of kind concepts (represented by the ovals) that FORK, LION, NECKLACE, and BOOK belong to. *Source:* Author.

FORK, additionally provides a perspective in which we understand there are principled differences in some of the properties of two types of (male and female) instances. More generally, the concept for any animate kind that reproduces sexually provides a perspective in which there are some principled differences between two types of instances (genders) of the kind, while the differences within each gender are understood to be accidental. This perspective is absent from the concepts for all inanimate kinds.[1]

These observations illustrate the variety and intricacy of the overlapping and distinguishing perspectives that are provided by DOG, NOTEBOOK, NECKLACE, and BOOK. The complex pattern of overlapping but distinct perspectives provided by the classes of kind concepts that these four concepts belong to is summarized in figure 2.1.[2] Those perspectives are further distinguished from those provided by the distinct classes of kind concepts that POEM, OBITUARY, ANIMAL, ROOK, WOOD, STORY, and BOOK each belong to. We will discuss the differing perspectives provided by the classes of concepts that those concepts belong to in chapter 5.

2.3 The Generality of the Perspectives Provided by Commonsense Concepts

The forgoing used individual concepts to illustrate the variety of intricate and abstract perspectives provided by kind concepts. It is important to note, however, that although individual concepts were used for the purposes of illustration, the perspectives provided by any individual concept are not uniquely provided by that concept, but rather are provided by a whole class of kind concepts of which the concept under consideration is a member. Thus, the perspectives provided by LION are also provided by CAT, HORSE, and many other concepts; those provided by FORK are also provided by PAD, PAPER CLIP, and many others; those provided by NECKLACE are also provided by DRESS, EARRING, and many others; and so on. It thus appears that the perspectives provided by commonsense concepts are intrinsically general. Although they may sometimes be identified in a single concept, there is nothing that requires those perspectives to be provided only by that concept. It is always possible for there to be another concept that provides the same perspectives. This suggests that perspectives are always distinguishable from the content of the concept, and thus potentially applicable to the content of other concepts within the same class.

2.4 Puzzles Raised by the Perspectival Nature of Commonsense Concepts

The perspectives provided by lexically expressible concepts are not properties of the things that the concepts represent. Instead, these perspectives are ways of thinking and talking about things and their properties that the concepts make available. The characteristics of the perspectives are characteristics of the "mental lenses" through which we think and speak about things and their properties, not properties of the things that the concepts represent. This raises a number of puzzles, which are addressed in part or full by the theory of conceptual form discussed in this book. These puzzles include the following:

1. How are perspectives represented? Existing approaches to conceptual representation have generally sought to represent concepts in terms of select properties of the things that they represent, or in terms of their conceptual roles. As such, these approaches have not addressed the question of

how the perspective that a concept provides may be represented. Insofar as commonsense concepts provide perspectives from which to think and talk, a theory of conceptual representation must address the question of how perspectives are represented.[3]

2. How do we acquire the perspectives? If perspectives do not represent the properties of things they allow us to think and talk about, but instead specify how we should think about things and their properties, then we obviously cannot acquire perspectives by simply observing the characteristics of the things that a concept allows us to think and talk about. The acquisition problem is further exacerbated by the normative and explanatory dimensions of the perspectives provided by commonsense concepts. It is easy enough to observe that many dogs have four legs, but how do we come to think that dogs are *supposed* to have four legs, and those that don't have something wrong with them? Similarly, it is easy enough to see that a given dog has four legs, but how do we come to understand that he has four legs because he is a dog? The variety and intricacy of the perspectives provided by commonsense concepts also complicate the learning problem, as the learner must have a way of figuring out which set of perspectives a given concept provides. Finally, the generality of perspectives raises the question of how a learner comes to know which other concepts provide the perspectives provided by a given concept. These questions are central to any plausible theory of concept acquisition. However, such questions have barely been recognized, much less addressed, because they come into focus only once the perspectival nature of commonsense concepts has been recognized and adequately characterized.

3. How do children know which perspectives should be mapped onto a given word?

Even if children do not have to acquire the perspectives provided by lexically expressible concepts because they are all innately given, they face a perspective identification problem. Children must have some way of knowing which set of perspectives a given word should be mapped onto. Studies show that from an early age, children map words used to refer to concrete objects onto concepts that provide an instance-of-object kind perspective, but this bias does not distinguish among the concepts that underlie words like "lion," "fork," "necklace," and "book," for example, all of which provide distinct additional perspectives. As such, children ultimately must have additional sources of information to determine which additional perspectives are provided by the concepts that a word is mapped onto. What are those sources, and how do they help identify the relevant perspectives?

4. How do children learn the perspective selection cues for selecting among the perspectives provided by a given concept? Given that lexically expressible concepts can provide multiple perspectives from which to think and talk, children must learn which cues select for one or another perspective provided by a given concept. For example, how do children come to know that "The dog" should be interpreted as being used to talk about an instance of the kind in "The dog was barking yesterday," but to talk about a kind in "The dog evolved from the wolf"?

5. Insofar as concepts in animals are not perspectival in this way, how did they evolve? Research in animal cognition has shown that animals are capable of extremely sophisticated cognitive feats. However, as far as I know, there is no evidence that their ability to achieve such feats uses mental representations that provide the types of intricate and abstract perspectives provided by even the simplest lexically expressible concepts. For example, although there is ample evidence that many animals can form concepts of objects and categorize objects, there is no evidence of which I am aware that they do so by representing an object as one of an unlimited number of objects that are the same in kind but differ numerically. Or that in doing so, they also think that some of the properties of that object are explainable by reference to the kind of thing it is, or that those are properties that the object is supposed to have by virtue of being that kind of thing, and objects without that property are not merely atypical but have something wrong with them. Nor, as far as we know, are animals capable of thinking about the kinds to which the objects belong as such (e.g., that dogs evolved from wolves). As such, we need an account of how concepts that provide multiple abstract perspectives of this sort evolved. The variety of such perspectives further complicates the evolution problem, as an account of how the large variety of perspectival concepts evolved is needed.

2.5 How Perspectives Are Represented: A Proposal

Traditional approaches to conceptual representation have focused on the content of conceptual representations rather than the form of conceptual representations. The inattention to form suggests that researchers have implicitly assumed that there is an arbitrary relation between the form of a conceptual representation and its content, just as there is an arbitrary relation between the phonological form of a word and its meaning.

In contrast, I propose that concepts possess formal structure (conceptual form) that is systematically related to its meaning. Just as the formal structure of a sentence provides instructions for how to interpret the relations between the contentful parts of a sentence, I propose that the formal structure of a concept provides instructions for how to interpret the relations between the contentful elements of a conceptual representation.[4]

We can get a better understanding of this proposal by first briefly considering how the formal (i.e., syntactic) structure of a sentence is related to the sentence's interpretation. Consider the sentence "Dogs bite people." The syntactic structure of the sentence provides instructions about how to think about the relation between dogs, people, and biting. It provides a perspective for thinking about dogs, biting, and people in which it is the dogs that do the biting and the people who get bitten. In contrast, the formal structure of the sentence "People bite dogs" provides a perspective in which it is the people who do the biting and the dogs that get bitten. Relatedly, we understand the sentence "The woman saw the girl with the binoculars" to be ambiguous—we don't know whether the girl or the woman has the binoculars—because there are two distinct syntactic structures that can be assigned to the sentence. Each syntactic structure provides different instructions as to how to think about the relations among the woman, the girl, binoculars, and seeing. These observations highlight the fact that the formal (syntactic) structure of sentences provide instructions for how to interpret how the contentful elements of the sentence are related to one another, and that in doing so, they provide a perspective from which to think about the relations among the contentful elements of the sentence.

Analogously, the perspectives provided by lexically expressible concepts may be represented via the formal structure of lexically expressible concepts—assuming that they have such a structure. Do lexically expressible concepts have a formal structure? Their perspectival nature strongly suggests they do. Furthermore, chapters 4 and 5 provide a detailed specification of the formal structure of kind concepts that provide the means for thinking and talking about the concrete objects and stuffs that we encounter. But before considering the formal structure of different classes of kind concepts, it will be useful to delineate the key ways in which the syntactic form of sentences and the conceptual form of lexically expressible concepts are similar, and how they differ, in the perspectives they provide.

2.6 Perspectival Properties of the Syntactic Form of Sentences and of the Conceptual Form of Lexically Expressible Concepts

Both syntactic form and conceptual form provide instructions for how to think about how their contentful elements are related, and thus provide a perspective from which to think about the contentful elements in relation to one another. Furthermore, the perspectives that they provide are abstract, in that they specify how to think about how the contentful elements are related, and the relations need not be ones that can be perceived.

The perspectives provided by both syntactic form and conceptual form are general, in that they do not relate a unique set of contentful elements but rather contentful elements that belong to classes of various types. Thus, the perspective provided by the syntactic form of "Scared white mice scamper quickly" is the perspective from which we interpret "Colorless green ideas sleep furiously," and indefinitely many other sentences that have the same syntactic form. Similarly, the perspectives provided by DOG are the same ones provided by CAT, HORSE, and indefinitely many other concepts that have the same conceptual form.

Both syntactic form and conceptual form also display a great variety. Furthermore, the variety of perspectives provided by the syntactic structures of sentences and the variety of perspectives provided by conceptual form do not vary arbitrarily. Research in linguistics over the past several decades shows that the syntactic structures found in the languages of the world are sharply constrained. There is unity in the diversity found in the syntactic structures found across languages. The constrained variety of syntactic structures found in the world's languages is generally understood to be the product of a generative system that allows variation in a small set of parameters (Baker, 2008).

Research on conceptual form is still in its infancy, and so one must be cautious about any conclusions concerning its nature, but the available evidence suggests that there is unity in the diversity found in the perspectives provided by lexically expressible concepts. In fact, it appears that a broad range of the perspectives provided by lexically expressible kind concepts can be generated by differences in only a small set of parameters. A detailed sketch of the generative system that yields the perspectives of a broad range of kind concepts is outlined in chapter 5.

In addition to those similarities between conceptual and syntactic form in the perspectives they provide, there are at least two important differences.

First, because of the recursive nature of syntactic structure, there is no limit to the number of different syntactic structures we can generate (in principle). But it is likely that there is a finite set of perspectives that are provided by lexically expressible concepts.

Second, and more important, conceptual and syntactic form differ in the nature of the instructions that they provide for interpreting the relations among the contentful elements. The conceptual form of kind concepts specifies intrinsic relations among the kinds, instances of kinds, and properties they have by virtue of being those kinds of things. Principled connections between a kind and a set of properties provide instructions to think of the properties as being lawfully connected to the kind, to have normative expectations concerning the property, and to think of the properties as being explained by the thing being the kind of thing that it is. As such, conceptual form specifies the intrinsic relations among aspects of things that constitute the nonaccidental systematic and principled unities we conceive of as existing in kinds and their instances. These relations are intelligible to us and guide how we think, reason, and talk about kinds, instances, and their properties. Conceptual form differs from syntactic form in all these respects.

The instructions provided by syntactic form do not require us to think of the contentful elements of a sentence as being lawfully connected to one another, bearing normative relations, or standing in explanatory relations to one another. Syntactic form specifies how to think that various aspects of the world are related to one another even if those aspects are not intrinsically, or actually, related in the specified manner. Syntactic form specifies how to think that elements are related to one another whether they are so related or not. As such, syntax provides the means for generating perspectives in which elements are not intrinsically related to one another and may not be actually related at all outside our thoughts. In contrast, conceptual form provides the means for thinking about the elements that are understood to be intrinsically related to what the concept represents.[5]

2.7 Toward a Solution to the Puzzles Raised by the Perspectival Nature of Commonsense Concepts

My proposal that lexically expressible concepts provide perspectives via the instructions for interpretation that are encoded in their formal structure, and a generative system yields the constrained variety of perspectives provided by lexically expressible concepts, may be dubbed a generative theory of conceptual form. Such a theory helps ameliorate but not solve the puzzles raised by the perspectival nature of commonsense concepts.

If there is a generative system that yields the perspectives provided by lexically expressible concepts, the system could constrain the perspectives that a learner considers when forming a concept, greatly simplifying the problem of how perspectives are acquired when we form concepts. Similarly, the perspective identification problem of how children learn which set of perspectives a given word is mapped onto is simplified if it can be constrained by the options made available by the generative system of conceptual form. Although early word-learning biases could potentially allow children to map words used to refer to concrete objects onto concepts that provide an instance-of-object kind perspective (Soja, Carey, & Spelke, 1991; LaTourette, Chan, & Waxman, 2023), such word-learning biases would not allow the child to know which further perspectives are provided by the concept for that kind of object. A generative system that yields the perspectives provided by lexically expressible concepts would provide the learner with a constrained hypothesis space for determining which additional perspectives are plausible when learning a new word.

Learning the linguistic and nonlinguistic cues that select among the multiple perspectives provided by a concept is also made easier by the constraints on perspectives provided by the generative system of conceptual form. Knowing, for example, that a given concept provides a perspective for thinking and talking about a kind as such, as well as a perspective for thinking and talking about instances of that kind, the learner could plausibly seek to determine which linguistic and nonlinguistic cues individually, or collectively, select for the kind or the instance-of-kind perspective.

Finally, the problem of how a large variety of concepts that provide abstract perspectives could have evolved, given that the perspectives do not simply track the properties of the things represented, is greatly simplified if the myriad perspectives are the product of a generative system. In that case, all (!) that is needed is an account of how the generative system evolved instead of an account of how a large number of distinct and unrelated perspectives evolved.

Clearly, then, there is much to be gained from investigating and characterizing the generative system that yields the perspectives provided by lexically expressible concepts. It is to that task that we turn in chapters 4 and 5.

2.8 Situating the Perspectives Provided by Lexically Expressible Concepts

Although the focus of this book is on the perspectives, or mental lenses, provided by lexically expressible concepts, we will also briefly consider conceptual combinations that provide the means for thinking and talking about contextually relevant ad hoc instances of kinds, ad hoc subkinds, and ad hoc kinds, as well as how the perspectives they provide are similar to, and different from, those provided by lexically expressible concepts in chapter 5. As we will see, although conceptual combinations can function in ways that are similar to the perspectives provided by lexically expressible concepts in some respects, there are also important differences.

Furthermore, as discussed previously, there are systematic differences between the perspectives provided by lexically expressible concepts and the perspectives provided by phrases and sentences more generally. Although they differ in the perspectives that they make available, both the conceptual form of lexically expressible concepts and the syntactic structure of a sentence provide instructions for interpretation and can be understood as functioning as "frames" in the sense of Camp (2019). According to Camp (2019, 311), a frame "constitutes a unified interpretive principle that organizes the characterizations to which it applies into more coherent and stable wholes" and function as "as over-arching principles for selecting, classifying, and connecting a subject's features into a multidimensional, intuitive cognitive structure." Camp recognizes many species of frames, including metaphors, analogies, just-so stories, and various elements of models in scientific theorizing. A theory of conceptual form may thus be an important addition to our understanding of the range of frames found in cognition.

One way in which the perspectives provided by lexically expressible concepts may turn out to be distinctive is that the principles that provide a unified interpretive principle are intrinsic to the concepts. In chapter 8, I outline one way in which the conceptual form of a specific lexically expressible kind concept (e.g., SHARK) can be used to interpret something other than things of that sort (e.g., LAWYERS) by investigating how conceptual form can guide the interpretation of metaphors. In metaphors, the perspective provided by one concept is used to interpret another concept. This suggests the possibility that conceptual form may also be presupposed by frames other than metaphor that are countenanced by Camp. However, specifying the relation between the conceptual form of lexically expressible concepts and other frames must await a specification of explicit theories about each of those frames.

II Conceptual Form of Kind Representations

3 The Instance-of-Kind Perspective and Limits of Existing Approaches

Although we only encounter particular things, we naturally think and talk about the particular things that we encounter as instances of one or another kind—as a dog, as a table, as a tree. It's an unremarkable observation, no doubt, but the cognitive feat that it reports—the naturalness of the instance-of-kind perspective—is indeed remarkable. To see why, it will be useful to look more closely at what is involved in taking the instance-of-kind perspective. As discussed in chapter 2, the instance-of-kind perspective, like other perspectives provided by lexically expressible concepts, is abstract. The instance-of-kind perspective is abstract in that it represents

- Instances as simultaneously unique and utterly nonunique within a kind
- A *member of* relation between each instance and the kind
- The things that we think and talk about as being intrinsically typed
- An explanatory link between the kind of thing that something is and some of its properties
- Normative expectations regarding some of an instance's properties
- The way that instances of the kind are expected to be similar without making similarity the basis of categorization

These are core facts about how we think and talk about things when we think of them as instances of a kind. Although this list is not exhaustive, it captures core, virtually undeniable facts about how we think of things when we think of them as instances of a given kind.

Section 3.1 explicates these six abstract characteristics of the instance-of-kind perspective. Section 3.2 shows why existing approaches to conceptual representation do not account for these characteristics of the instance-of-kind perspective. Finally, section 3.3 argues for the need for a new approach to representing kind concepts—one that can capture the core characteristics of the instance-of-kind perspective.

3.1 Unpacking the Abstract Characteristics of the Instance-of-Kind Perspective

Consider a situation in which I encounter Jett and think of him as a dog. In thinking of Jett as a dog, I think of him in a way that allows me to simultaneously think of him as unique and utterly nonunique. He is unique in that he is a different dog than any and all other dogs. He is different from all actually existing dogs, as well as potentially existing dogs. Quite simply, no other dog is, or can ever be, Jett, and Jett cannot be any dog other than the dog he is.

At the same time, Jett is utterly nonunique, in that he is the same kind of thing as an unlimited number of other actually and potentially existing dogs. I do not need to explicitly or consciously entertain these propositions to think of Jett as a dog. Instead, I use a representation that has a form that represents him as simultaneously being unique and nonunique within the category of dogs. That representation allows us to think and talk about Jett as a dog, as well as to form explicit thoughts about his uniqueness and nonuniqueness within the category of dogs. Chapter 4 contains a proposal concerning the form of such a representation.

It is notable that to think and talk of Jett as being the same kind of thing as an unlimited number of other dogs, I do not need to know whether these non-Jett dogs actually exist. In fact, I could maintain this perspective even while believing that no dogs other than Jett exist! Simply being able to think of a single instance, Jett, as a dog is all that is needed to think of him as simultaneously being unique and utterly nonunique within the kind DOG. As such, *the instance-of-kind perspective provides a way of representing a particular as simultaneously unique and utterly nonunique within a kind.*

In addition to making available the simultaneous uniqueness and nonuniqueness relations among instances of a kind, the instance-of-kind perspective makes available the *member-of* relation that instances have to kinds. In thinking of Jett as a dog, I think of him as member of the kind DOG, where the kind DOG is an abstract, setlike entity that contains an unlimited number of dogs, including Jett. Certainly, none of us has ever encountered this abstract entity. Nevertheless, we easily think and talk about it. For example, when we think or say that dogs evolved from wolves, or that dogs are not on the verge of extinction, we are thinking and talking about a kind rather than properties of members of the kinds. On the other hand, statements like "Jett is a dog" are understood as identifying Jett as a member of the kind DOG. As such, *the instance-of-kind perspective provides a way of representing a member-of relation between an unlimited number of instances and an abstract kind.*

When we encounter something unfamiliar and do not know what kind of thing it is, we want to know what it is. "What is that?" is about the most basic question that we can ask about something. Furthermore, we expect the question to be answered by identifying the thing as an instance of a kind (e.g., "That is a dog/table/piece of wood/ . . ."). An answer professing ignorance concerning the kind of thing that something is is also acceptable. In such cases, we expect there will be experts who know the answer. What does not seem to be an acceptable answer is one saying that there is no fact of the matter as to what kind of thing something is, or that it is no kind of thing at all.[1]

Although perception and action use untyped representations to represent the properties that a given thing has at any given moment, and thus to dynamically adjust to changes in any property of a thing (e.g., Kahneman, Treisman, & Gibbs, 1992; Leslie, Xu, Tremoulet, & Scholl, 1998; Pylyshyn, 1989, 2001; Yu & Lau, 2023), thinking, talking, and reasoning require us to identify what kind of thing something is. It is only once we identify something as an instance of a kind that we know how to think about its properties—which ones are accidental, essential, necessary, etc. Untyped representations cannot make such distinctions among the properties of a thing. All properties that are bound to such representations are contingent, and all that can be known is that they are true of the thing at a given time. This type of representation is advantageous for perceptual and action systems that must dynamically adjust to the information that they have about a given thing at a given time. Untyped representations can also be useful in working memory, where we are concerned only with temporary representations of things (Luck & Vogel, 1997). Thinking and reasoning, on the other hand, require knowing what kind of thing we are thinking about because only then can we know which properties we can rationally assume to be true of instances of a kind in the absence of the moment-to-moment information provided by our perceptual systems. As such, our conceptual systems require the things that we think and reason about to be represented as instances of one kind or another. For the conceptual system, there is nothing that is not some kind of thing.

And yet, despite that, most theories of categorization (wrongly, in my opinion) implicitly or explicitly assume that our conceptual systems contain representations of untyped particulars, often called "token representations," and that categorization is the result of *combining* such an untyped token representation with a type representation via predication or some other mechanism (e.g., Jackendoff, 1983; Barsalou, 2012). But if we accept that way of understanding categorization, particulars that are of no specific type should be common, as there will be many circumstances in which we have token representations of a particular that we encounter and do not know what type of thing it is (and thus no type representation is predicated of the token representation).

But commonsense conception appears to reject the idea that the things that we encounter are things of no specific type. In accordance with the instance-of-kind perspective, *the things that we think and speak about are understood to be intrinsically typed* (even if we don't know what that type is).

Whereas it has traditionally been thought that categorizing something as an instance of a kind is a distinct cognitive process from understanding or explaining why that thing has the properties that it does, it is now evident that thinking of something as an instance of a kind not only categorizes the thing but also renders some of its properties intelligible. Categorization and explanation of properties are intrinsically tied together (e.g., Carey, 1985; Murphy & Medin, 1985; Keil, 1989; Wellman & Gelman, 1992; Gopnik & Meltzoff, 1997).

So, for example, if I think of something as a table, I understand it to have a nonarbitrary structure (Prasada, Ferenz, & Haskell, 2002), and thus one may explain its shape as being what it is because it is the kind of thing it is. On the other hand, if I think of that same thing as a hunk of wood, I think of its shape as being arbitrary (Prasada et al., 2002), and thus one could not explain its shape as being due to its being wood. As such, *the instance-of-kind perspective provides an explanatory link between the kind of thing something is and some of its properties*. This link supports a form of explanation, called "formal explanation," in which the kind of thing something is is used to explain certain properties of the thing (e.g., Jett has four legs *because* he is a dog). For all kinds, the instance-of-kind perspective supports formal explanations of some of the properties of instances of the kind (Prasada & Dilllingham, 2006, 2009; Prasada, 2017; Rivera, Prasad, & Prasada, 2023). Depending on the kind of thing that we are thinking about, the instance-of-kind perspective may also contain additional structure that supports causal and/or teleological explanations (e.g., Carey, 1985; Murphy & Medin, 1985; Keil, 1989; Kelemen, 1999; Lombrozo, 2009; Strevens, 2000).

When we think of things as instances of kinds, we understand certain properties of the thing to be properties that they are *supposed* to have, and if an instance lacks that property, it is viewed as having something *wrong* with it. That is, the instance-of-kind perspective provides normative expectations concerning those properties. For example, we think that dogs are supposed to have four legs. If they don't have four legs, they are thought to have something wrong with them. This tells us that *thinking of something as an instance of a kind not only categorizes a thing, it also generates normative expectations regarding some of its properties*. That is quite remarkable—experience only provides us with evidence concerning what *is* the case about particulars, and yet the instance-of-kind perspective provides normative expectations such that an instance of a given kind is supposed to have certain properties simply because it is that kind of thing.

Finally, the instance-of-kind perspective allows for the possibility that something could be highly similar to instances of a kind and yet not be thought to be an instance of that kind. For example, although dingos are very similar to dogs generally, we don't think of them as dogs. Furthermore, the instance-of kind perspective also allows that something that is quite dissimilar from category members generally may nevertheless be thought to be an instance of the kind (Gelman, 2003). For example, penguins are quite dissimilar from most birds, but penguins are nevertheless thought to be birds. Such observations may suggest that the instance-of-kind perspective is unconnected to the similarity of the instances in a kind, but this is not the case. Quite the contrary, in fact—the kind of thing that something is determines the ways in which instances of that kind are generally expected to be similar to one another. This is because instances of a kind are understood to have certain properties because they are the kinds of things that they are. These are properties that we think instances of that kind are supposed to have. As such, we expect instances of the kind to generally have these properties and thus be similar in these respects. *The instance-of-kind perspective specifies how instances of the kind are expected to be similar without making similarity the basis of categorization.*

These observations show that although the act of thinking of something as an instance of a kind is natural and experienced as unremarkable, it is in fact a complex cognitive feat that involves many abstract and nonobvious characteristics.

3.2 Existing Approaches to Representing Kinds and the Instance-of-Kind Perspective

Any theory of kind concepts must provide an account of the instance-of-kind perspective, and thus it must account for the six core characteristics of the perspective outlined here. Existing theories of conceptual representation and generic knowledge, which represent kinds via summary descriptions and sets, respectively, fail to account for these core characteristics of the instance-of-kind perspective. In some cases, it is possible to think of an addition or modification to the approaches that would allow them to capture one of the core characteristics, but in no case is the modified approach able to capture the other core characteristics of the instance-of-kind perspective. Furthermore, there is no way to modify the approaches to capture all (or even most of) the core characteristics of the instance-of-kind perspective. Finally, framework theories presuppose kind representations and thus do not themselves capture characteristics of the instance-of-kind perspective. As such, a new approach to representing kind representations is needed and is outlined in this book beginning in chapter 4. But before we spell out this new approach, it is useful to examine which characteristics of the instance-of-kind perspective are captured by previous research, as well as the reasons why they fail to capture other aspects.

3.2.1 Representing Categories via Summary Descriptions

Research on conceptual representations generally assumes that concepts that represent categories do so via summary descriptions of category members (Murphy, 2002).[2] Approaches differ, however, in the nature of the summary representations that they propose. The classical theory proposes that the descriptions take the form of definitions. Definitions specify the properties that are understood to be individually necessary and jointly sufficient for being a member of a category. These properties are understood to be shared by all members of the category. The prototype approach, on the other hand, proposes that categories are represented via prototypes (Rosch, 1975). Prototypes are descriptions that contain properties that are typical of category members. The properties in the prototype are understood to be shared by many, although not necessarily all, members of the category. Finally, some taking the explanation-based approach propose that categories are represented via a description of the causal-explanatory structure that is shared by category members. In that case, the summary descriptions contain not only category-typical properties, but also the causes that explain those properties; thus the summary descriptions represent the causal explanatory structure that is shared by category members.

These differences in the nature of the summary descriptions posited by the various approaches to conceptual representation have important consequences for how we think and talk about things. However, we will focus on some shared strengths and limitations of summary descriptions in capturing the abstract characteristics of the instance-of-kind perspective discussed thus far. Differences between approaches to conceptual representation will be discussed only insofar as the approaches diverge in their ability to capture any of the six abstract characteristics of the instance-of-kind perspective under consideration.

3.2.1.1 Simultaneous Uniqueness and Nonuniqueness of Instances Within a Kind Because all the approaches described here represent categories via a summary representation of *what is shared* by members of the category, these theories can represent the respects in which something that is thought of as an instance of a kind is utterly nonunique within the category. For exactly this reason, these conceptual representations cannot represent the uniqueness of anything that is thought of as an instance of the kind. Quite simply, the properties that instances of a kind share cannot be used to distinguish individual instances. This limitation helps highlight the fact that if kind concepts are represented via summary descriptions, concepts cannot, on their own, provide an instance-of-kind perspective that allows us to think of an instance of a kind to be simultaneously unique and utterly nonunique within the kind. Summary descriptions only represent what is common to category members, so they can only represent the respect in which an instance of a kind is utterly nonunique within that kind.

To think of something as an instance of a kind, summary descriptions must at minimum be combined with a representation that provides information that distinguishes that instance from other instances of the kind. One way to do this would be to combine a summary description of a category/kind and a representation of an untyped particular.[3] If the representation of each untyped particular is (or contains) a unique indexical representation that represents a thing and distinguishes it from other things without describing it, the combined representation will be able to represent the uniqueness of the thing represented by the combined representation. In the combined representation, the properties that are in the summary description, and thus properties that are shared by members of the category or kind, are attributed to the particular. Although summary descriptions do not provide an instance-of-kind perspective, the combined representation of a summary description and an indexical representation of an untyped particular can capture a key element of the instance-of-kind perspective, representing an instance as simultaneously unique and utterly nonunique within the kind.

3.2.1.2 Member-of Relation Between Instances and Kinds Do summary descriptions, or the combined representation described here, make available the member-of relation between instances and kinds that is provided by the instance-of–kind perspective? A moment's reflection reveals that they do not. We think and speak of kinds as abstract entities that contain an unlimited number of members.

The combined representation characterizes a single particular that may represent a member of a kind, but it cannot represent the kind itself. The summary description, being a description in terms of the properties that are shared by category members, is not characterized in terms of quantity. A description may apply to any number of things, but those things are not contained within the description.

Kinds, on the other hand, essentially contain an unlimited number of instances. A given kind may, at different times, contain different members and different numbers of members. Descriptions never contain any members. The things that a description applies to cannot be thought to be members of the description, although they may be thought to be members of a kind. Summary descriptions that represent categories may characterize a kind and many of its members but they cannot represent the kind itself, as it is a description that does not represent quantity, and an essential component of a kind representation is that it must be able to represent the fact that it contains an unlimited number of instances.

As such, approaches to conceptual representations that represent categories via a summary description of the properties shared by category members cannot represent the kind itself, and thus they also cannot represent a member-of relation between an unlimited number of instances and a kind.

3.2.1.3 Intrinsic Typing of Particulars As discussed previously, in thinking of things as instances of a kind, we think of the particular things that we think and talk about as being intrinsically typed. Concepts that represent categories via summary descriptions do not account for this characteristic of the instance-of-kind perspective. The reason is that the summary descriptions specify what is common to members of a category; they do not specify particular instances. It is usually assumed that the summary descriptions apply to representations of particulars, and in doing so, specify the type of the particular. As such, theories that represent categories via summary descriptions assume the existence of representations of untyped particulars and cannot account for the fact that we resist the idea that the things we encounter are no kind of thing (even when we do not know what type of thing something is).

3.2.1.4 Explanatory Link That Supports Formal Explanation The instance-of-kind perspective provides an explanatory link between the kind of thing that something is and some of its properties. This link supports formal explanations of those properties—explanations of the properties by reference to the kind of thing that it is (e.g., Jett has four legs because he is a dog). The classical and prototype approaches to conceptual representation do not capture this characteristic of the instance-of-kind perspective, as they do not represent explanatory structure generally, much less the explanatory structure relevant for formal explanation.

Explanation-based approaches that posit summary descriptions represent the properties that are characteristic of the kind and their causes. In so doing, they represent explanatory links between properties shared by members of the category, as well as their (potentially unobserved or unknown) causes. To represent an explanatory link between the kind of thing that something is and the properties for which formal explanations are possible, the summary descriptions would have to represent the kind. But, as discussed earlier, summary descriptions describe what members of a kind are like but do not provide a representation of the kind itself.

As such, the summary descriptions posited by explanation-based approaches cannot represent the type of explanatory structure needed to support the formal explanations that are made available by the instance-of-kind perspective, as that requires identifying the kind as what explains why instances of that kind have certain properties.

3.2.1.5 Normative Expectations The instance-of-kind perspective generates normative expectations regarding certain properties that characterize the kind. Theories that represent categories as summary descriptions typically represent the properties (and explanatory structure, in the case of the theory approach) that describe category members. However, they can easily be modified to represent the properties that we think category members are *supposed to* share rather than the properties that they *do* share. In fact, the prototype approach sometimes takes the prototype to represent the ideal category member (Barsalou, 1985; Lynch, Coley, & Medin, 2000; Burnett, Medin, Ross, & Blok, 2005; Foster-Hanson & Rhodes, 2019). Representing the properties of the ideal member of a category involves a different form of normativity than representing the properties that category members are supposed to have. Nevertheless, the existence of such proposals shows that the prototype approach can potentially incorporate the normative expectations generated by the instance-of-kind perspective. But such a change would raise the question as to how summary representations containing normative expectations arise. If category representations are summary descriptions of category members, the summaries can be derived from descriptions of the properties of individual category members. If, on the other hand, the summaries contain normative expectations concerning the properties that are supposed to be shared by category members, there needs to be a mechanism by which we can go from descriptions of the properties of individual members of a category to normative expectations regarding the properties that they are *supposed to* have. To date, the prototype approach does not provide such a mechanism.

Theories that represent categories via summary descriptions of the shared causal-explanatory structure of a category could also be amended to generate the types of normative expectations generated by the instance-of-kind perspective. Instead of representing the properties that category members share and their causes, the amended theories would represent the properties that category members are supposed to have and their causes. But incorporating normative expectations in the summary descriptions in this manner would raise the same question faced by the prototype approach: Given only knowledge of the properties that members of a category do have, how do we come to conclusions concerning what properties they are *supposed* to have?

Whereas it is possible to envision how prototype and explanation-based summary descriptions may be amended to incorporate the normative expectations generated by the instance-of-kind perspective, the classical approach leaves no room to do so. Doing so would require representing the properties that are part of the summary representation as not being individually necessary for category membership. This would contradict the classical approach's idea that categories are represented in terms of definitions that specify the properties that are individually necessary and jointly sufficient for category membership.

3.2.1.6 Similarity Relevance Without Similarity Dependence Finally, the instance-of-kind perspective specifies how instances of the kind are expected to be similar without making similarity the basis of categorization. This characteristic of the instance-of-kind perspective is easily captured by the summary representations posited by the theory-based approach because that approach recognizes two levels of representation—the properties that are characteristic of the kind and shared generally by category members and the causes of those properties (Medin, 1989; Gelman, 2004). By representing the properties that are characteristic of the kind, it represents the ways in which category members are expected to be similar. Because it also represents the causes of those properties, it allows for the possibility that two things could have similar properties but different causes of those properties, and thus they could be members of different categories. It also allows for the possibility that the same causal structure may be realized in different ways, and thus that two things could be quite dissimilar and yet be members of the same category. As such, the summary descriptions that are posited by the theory-based approach for representing categories can capture the characteristic of the instance-of-kind perspective, whereby it specifies how instances of the kind are expected to be similar without making similarity the basis of categorization. Because the classical and prototype approaches posit only a single level of representation (Medin, 1989), they cannot capture this characteristic of instance-of-kind representations.

3.2.2 Representing Categories as Sets

Having examined which characteristics of the instance-of-kind perspective are and are not captured by existing approaches to conceptual representations that represent categories via summary descriptions, we now will turn to considering approaches assuming that kinds can be represented via sets.

Within the concepts literature in psychology, the classical theory views categories as sets. However, according to this theory, categories are not explicitly represented. Instead, we represent categories via definitions, which are summary descriptions that specify the conditions for category membership. As discussed previously, summary descriptions fail to capture many characteristics of the instance-of-kind perspective.

Within linguistics, much work on generic sentences such as "Dogs bark," which are used to talk about kinds of things, posits that categories and kinds can be represented as sets (Carlson & Pelletier, 1995). Although researchers vary in whether they think that sets are mentally represented, it will be useful to examine which characteristics of the instance-of-kind perspective can be captured by set representations.

3.2.2.1 Simultaneous Uniqueness and Nonuniqueness of Instances Within a Kind As has been discussed, the instance-of-kind perspective represents instances as simultaneously unique and utterly nonunique within a kind. This is also how members of a set are represented. Each member is unique insofar as it is different from every other member of the set. At the same time, each member is utterly nonunique in that, like every other member of the set, it meets the conditions for being a member of that set. As such, set representations can capture the simultaneous uniqueness and nonuniqueness of members of a category that the instance-of-kind perspective provides.

3.2.2.2 Member-of Relation Between Instances and Kinds Similarly, just as the instance-of-kind perspective makes available a member-of relation between instances and a kind, set representations make available a member-of relation between elements and sets.

3.2.2.3 Intrinsic Typing of Particulars According to the instance-of-kind perspective, the things that we think and speak about are understood to be intrinsically typed. Set representations do not make this assumption. Elements exist and can be distinguished and identified independent of the sets that they belong to. As such, the elements do not intrinsically belong to any given set, and we can identify and talk about elements independent of the different sets that any given element belongs to. As such, set representations, unlike the instance-of-kind perspective, do not assume that particulars (i.e., elements) are intrinsically typed.

3.2.2.4 Explanatory Link That Supports Formal Explanations The instance-of-kind perspective provides an explanatory link between the kind of thing that something is and some of its properties. This link supports formal explanations of those properties. As such, classification of things and rendering some of their properties intelligible via formal explanation are intrinsically connected within the instance-of-kind perspective. This is not so for set representations, which classify without rendering any of the properties of the things classified intelligible via formal explanation (or any other type of explanation). As such, set representations do not capture this characteristic of the instance-of-kind perspective.

3.2.2.5 Normative Expectations Just as the classical approach to conceptual representation leaves no room for representing the category in terms of the properties that category members are supposed to have, set representations leave no room for the normative expectations generated by the instance-of-kind perspective. If something lacks a property that is required for set membership, it is simply not a member of the set. It cannot be represented as a defective member of the set. It is worth noting that adopting fuzzy set theory is of no help in this regard. Fuzzy set theory provides a way for set theory to incorporate typicality phenomena by making it possible for items to be members of a set, more or less (Zadeh, 1965). However, this approach cannot capture the normative expectations generated by the instance-of-kind perspective. The reason for this is that instances of a kind that lack a property that instances of that kind are supposed to have are not merely less typical of the category—they are judged to have something wrong with them. As such, the normative expectations introduce a normative element over and above differences in typicality that underlie graded membership. Atypical members of a category (e.g., chihuahuas for DOG) need not be thought as having anything wrong with them, but a typical member of a category missing a property that instances of the kind are supposed to have (e.g., a three-legged Labrador for DOG) is thought to have something wrong with it. It is thus clear that set representations cannot capture the normative expectations generated by the instance-of-kind perspective.

3.2.2.6 Similarity Relevance Without Similarity Dependence Finally, the instance-of-kind perspective specifies how instances of the kind are expected to be similar without making similarity the basis of categorization. As discussed earlier, this characteristic requires representing both the properties that are characteristic of a kind and their causes. Set membership, on the other hand, simply requires that members have the properties required for set membership without regard to what, if anything, caused those properties. As such, set representations do not capture this characteristic of the instance-of-kind perspective.

3.3 The Need for a New Approach

It is thus clear that despite its centrality to our cognitive lives, the instance-of-kind perspective cannot be represented via existing approaches to conceptual representation or set representation. Furthermore, many of their limitations appear to be fundamental in nature, and thus it does not appear that they can be modified to account for the core characteristics of the instance-of-kind perspective. The limitations of existing approaches serve as the motivation for developing a new approach to the representation of kind concepts—one that minimally captures the core characteristics of the instance-of-kind perspective. In the next two chapters, I propose a new approach, which accounts for the core characteristics of the instance-of-kind perspective and the perspectival properties of kind concepts more generally.

4 Mechanisms for Thinking About Kinds and Instances of Kinds

Existing approaches to conceptual representation that represent kinds via summary descriptions of what is *shared* by instances of the kind fail to account for the core characteristics of the perspective that we take when we think of something as an instance of a kind. Framework theories presuppose the ability to form kind representations and thus also do not provide an account of the instance-of-kind perspective. As such, a new approach to representing kind concepts is needed. The new approach must minimally make available the instance-of-kind perspective, as well as the kind perspective, because kind concepts not only allow us to think of things as instances of a kind, they allow us to think thoughts about kinds themselves (e.g., BALD EAGLES ARE ENDANGERED).

The new approach must overcome three limitations of summary descriptions that became evident when considering the core characteristics of the instance-of-kind perspective. First, because they represent kinds implicitly via the properties shared by instances of a kind, they cannot represent the explanatory link between the kind and certain properties of instances of the kind, or the member-of relation between the kind and its instances. This limitation also prevents summary descriptions from representing the kind perspective because the properties that can characterize the kind per se (e.g., extinct, widespread, invented) do not apply to instances and thus cannot be shared properties of instances. Second, because summary representations do not represent how instances are distinguished from one another, they are unable to represent the uniqueness of each instance of the kind or the different ways in which instances of kinds such as "dog," "water," and "book" differ from one another. Third, because they do not represent the quantitative dimension of kind concepts, summary descriptions are unable to capture the fact that each instance is understood to be one of an unlimited number of instances of that kind.

I turn now to developing a theory of how kind concepts are represented that overcomes these limitations.

4.1 Strategy for Developing a New Approach to Representing Kind Concepts

In this chapter, I will develop the formal structure of kind concepts that belong to the class of kind concepts that includes concepts such as DOG, CAT, and TREE. The kind concepts within this class embody the components in terms of which we can generate a broad range of formally distinct classes of kind concepts.

For ease of exposition, I will develop the representation of kind concepts in two stages. In the first stage, the core mechanism for representing kind concepts will be developed. To do so, I begin by identifying four essential characteristics of how we think of kinds and instances of kinds along both qualitative and quantitative dimensions. I will then specify the core mechanism for representing kind concepts by identifying the types of formal components implicated by each of those characteristics. Finally, I will illustrate how the proposed mechanism not only captures the essential facts that motivated the mechanism, but also captures the core characteristics of the instance-of-kind perspective discussed in chapter 3.

In the second stage, I will elaborate the mechanism by identifying and incorporating four further characteristics of how we think of kinds and instances of kinds, and show how the elaborated mechanism can account for further facts about how we think and talk about kinds and their instances.

4.2 Characterizing the Core Mechanism for Representing Kind Concepts

4.2.1 Four Essential Characteristics of How We Think and Talk About Kinds and Instances of Kinds

Here are four characteristics that any theory of the representation of kind concepts must capture.

(i) Kinds are understood to have an unlimited number of instances. In thinking and talking about a kind, we are thinking and talking about an abstract entity that is understood to have an unlimited number of instances in it. A statement that characterizes a kind, such as "Dogs bark," is not just about the dogs that happen to exist at the time of the utterance, but also dogs that existed previously, will exist in the future, exist in counterfactual scenarios, and other scenarios (Carlson, 1977; Macnamara, 1986; Murphy, 2002). The statement is about an unlimited number of instances of the kind. Furthermore, there is no limit on the number of instances of a kind we can think of. One can always think of another instance of the kind that differs from all other instances of the kind. The intuition that, in thinking and talking about a kind, we are thinking of an unlimited number of instances is supported by some recent experimental data. Fenton and Prasada (2024) presented participants with generic sentences that characterize kinds (e.g., "Dogs bark"/"Mosquitoes carry West Nile virus"/"Cars have radios") and nongeneric sentences that stated something about sets of instances of a kind (e.g., "Many dogs were barking"/"Five cars were in the parking lot") and asked them to rate the extent to which the sentences were saying something about (1) an unlimited number of instances of the kind named by the noun, (2) an unspecified but finite number of instances of the kind named by the noun, or (3) a specific number of instances of the kind named by the noun. Participants interpreted the generic sentences as being about an unlimited number of instances of the kind and interpreted nongeneric sentences to be about an unlimited number of instances only if they involved a universal quantifier such as "all," "every," or "each." These data support the intuition that kinds are understood to have an unlimited number of instances in them.

(ii) Instances of a kind only need to differ numerically. Instances of a kind (e.g., Fido and Rover) typically differ in their qualities, and thus we can often distinguish one instance of a kind from other instances of the same kind by the qualities that are unique to an instance. However, we do not require distinct instances of a kind to be qualitatively distinct. Instances of a kind can be qualitatively identical; they need only be numerically distinct. It is easier to imagine qualitatively identical but numerically distinct instances of some kinds (e.g., "paper clips") than others (e.g., "dogs"), but whether they actually exist or not, we can think of numerically distinct instances that are qualitatively identical. Furthermore, this is not a mere theoretical possibility—we think of instances as being numerically distinct while being qualitatively identical when we think of things as being duplicates or clones, and we also think of instances of mass-produced products as being numerically distinct while being qualitatively identical.

(iii) Kinds must differ qualitatively. Kinds, unlike instances of kinds, cannot be merely numerically distinct. Thus, it makes no sense, for example, to say that there are two kinds—"dogs" and "shmogs"—that are qualitatively identical but nevertheless numerically distinct. In such a circumstance, we are led to conclude that there is a single kind that has two names. Notice how different this is from how we think about instances of kinds. Given evidence of two qualitatively identical instances of some kind—Fido and Rover, for example—we may be surprised to find that the two instances are qualitatively identical and may search for some small qualitative difference that complements the numerical difference, but there's no urge to conclude that there is only one instance with two names. Told that dogs and schmogs are qualitatively identical, we demand evidence of some qualitative difference between the two that would justify saying that there are two kinds, not one.

Some empirical evidence that supports that intuition, as well as the intuition that instances of a kind may be merely numerically different without also being qualitatively distinct, comes from a recent experiment by Fenton and Prasada (2024). In the experiment, participants were presented with two scenarios. In one scenario, a speaker says that he saw a raccoon named Bob and, at another time, a raccoon named Dave, and that the two raccoons were identical. An interlocuter responds that there must be some unknown qualitative difference between the raccoons, and that if there isn't one, then the speaker saw one raccoon with two names. The other scenario had a parallel structure, except that the speaker tells of learning about two kinds of animals—raccoons and "schmaccoons." The interlocuter then responds that there must be some unknown qualitative difference between the raccoons and the schmacoons, and that if there isn't one, then the speaker learned about one kind of thing with two names. In both cases, participants were asked to rate how reasonable the interlocuter's response is. As predicted by the intuitions that instances of kinds need only be numerically distinct but kinds that are numerically distinct must also be qualitatively distinct, participants rated the response as more reasonable in the kind condition than the instance-of-kind condition.

(iv) The qualitative character of a kind is nonaccidental. Qualities that characterize a kind and thus distinguish it from other kinds are lawfully connected to the kind, as these properties are expected to be true of an unlimited number of instances of the kind. Qualities that distinguish between kinds must also, directly or indirectly, be lawfully connected to the other properties that characterize the kind and the kind itself, as these properties exist jointly in an unlimited number of instances of the kind. Together, they characterize the kind and distinguish it from other kinds. Instances of a kind, on the other hand, need not be distinguished qualitatively, and thus a quality that happens to distinguish two instances is understood to be accidentally related to the qualities that the instances of the kind share by virtue of being that kind of thing. That helps explain why we think that kinds differ from one another in principled ways but instances of a kind differ from one another in accidental ways.

4.2.2 Formal Mechanisms That Embody the Essential Characteristics of Kind Concepts

Given that there are an unlimited number of instances in any kind, representing the kind by enumerating its instances is not a viable approach. We could not possibly list all the instances of a kind at a given time, and even less so the instances that existed in the past or that could exist in counterfactual circumstances. Another possibility that one might consider is to represent kinds via a summary of the characteristics shared by instances of the kind. This is what existing approaches to conceptual representation do. As discussed previously, there are many ways in which that way of representing kinds is inadequate. A different way of representing kinds is needed.

What is needed is a generative mechanism (5) that can generate instance-of-kind representations (6) without limit. Because the mechanism can generate instance-of-kind representations without limit, it implicitly contains any and all instance-of-kind representations of that kind. Consequently, (5) can be used to think of the kind as such.

(5) K_i

(6) K_1 K_2 K_3 K_4 K_5...

How exactly does the generative mechanism that represents kinds (5) generate an unlimited number of instance-of-kind representations (6)? At this stage of our inquiry, I leave that unspecified. The key insight is that we need a mechanism with the formal characteristic that it has the capacity to generate an unlimited number of instance-of-kind representations. There are likely many ways of realizing that formal characteristic. In chapter 11,

I present one interesting possibility for what the generative mechanism for representing kinds may turn out to be. For the present purposes, though, all that is needed is the existence of a generative mechanism (5) that can generate an unlimited number of instance-of-kind representations of the form in (6), not the specific form of the mechanism.

Turning to the instance-of-kind representations (6) generated by the kind mechanism, we note that each instance-of-kind representation consists of two parts. One part is shared by each instance-of-kind representation generated by a given kind mechanism. It represents the kind of thing that the instance is understood to be. The other component is an index that is unique to each instance-of-kind representation generated by a given kind mechanism. This index represents the uniqueness of the instance within the kind. This component of instance-of-kind representations is an index and thus has no intrinsic descriptive content. As such, it provides a way of distinguishing instance-of-kind representations without assuming any qualitative difference between the instances. The instance-of-kind representations, and thus their indices, are produced by the generative mechanism for that kind, and therefore the indices have the function of numerically distinguishing between instances of that kind and identifying each instance as one instance of that kind.

It will be useful to represent these formal characteristics of kind concepts as in figure 4.1. The primary advantage of this representation is that it makes it easier to distinguish the distinct components of the conceptual form, and thus to see how the conceptual forms of one class of kind concepts are similar to and different from the conceptual forms of other classes of kind concepts, which we will be developing and discussing in chapter 5. A second advantage of this representation is that it serves as a reminder that the formal characteristics are formal characteristics of a *mechanism,* which is presumably neurally instantiated, that allows us to think and talk about kinds, instances, and subkinds. That is, the goal is to characterize the real mechanism that allows us to think and talk about kinds, instances, and subkinds and their properties.

The mechanisms represented in (5) and (6) may be represented in the following manner, with the generative mechanism for representing the kind depicted in the top part of figure 4.1 and the unlimited number of instance-of-kind mechanisms generated by the kind mechanism depicted in the lower part.

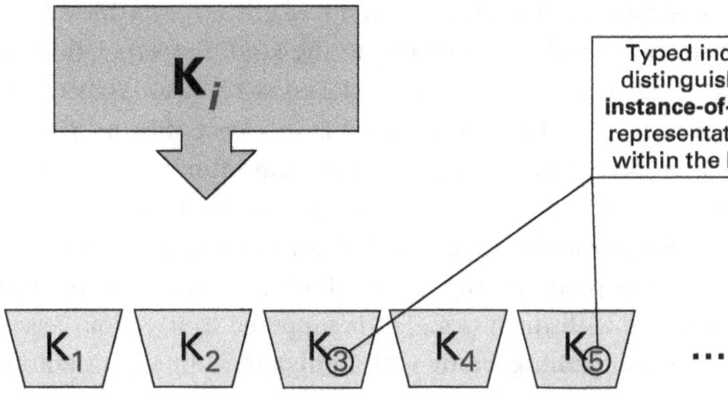

Figure 4.1
Depiction of the generative component of the mechanism for representing kinds. The generative mechanism (top) has the capacity to generate an unlimited number of instance-of-kind representations (bottom). *Source:* Author.

The representation developed thus far does not provide a way of distinguishing kinds. As previously discussed, kinds cannot be merely numerically distinct; they must also differ qualitatively. Furthermore, the properties that characterize a kind and distinguish it from other kinds must be lawfully connected to that kind and the other properties that do so. The question, then, is how to represent the qualities that are lawfully connected to a kind and characterize and distinguish the kind from other kinds. Clearly, different properties are relevant for different kinds, and the nature of the lawful connections may also vary across content domains. Furthermore, given that the theory is seeking to specify the *formal* structure of kind representations, it cannot represent the specific properties that characterize specific kinds. Instead, the formal structure of kind representations must include the *form of the connection* that properties that characterize and distinguish a kind from other kinds have to the kind.

Each kind must have a set of properties that it has by virtue of being that kind of thing. These properties characterize the kind and distinguish it from other kinds. Recent research shows that our conceptual systems distinguish properties that kinds have by virtue of their being the kinds of things they are from properties that are lawfully connected to the kind in other ways (Prasada & Dillingham, 2006, 2009; Prasada, Khemlani, Leslie, & Glucksberg, 2013; Haward, Wagner, Carey, & Prasada, 2018; Haward, Carey, & Prasada, 2021). Properties that are understood to be true of a kind by virtue of its being the kind of thing that it is (e.g., four-leggedness for dogs) are said to have a *principled connection* to the kind. Properties that have a principled connection to a kind are understood to be an aspect of being that kind of thing (e.g., being four-legged is an aspect of being a dog). Furthermore, a property that has a principled connection to a kind supports (1) a formal explanation of the property (e.g., Jett has four legs because he is a dog); (2) the normative expectation that instances of the kind are supposed to have the property, and if they don't have the property, there is something wrong with them (e.g., Jett is supposed to have four legs; if he doesn't, there is something wrong with him); and (3) the expectation that instances of the kind will generally have the property (e.g., in general, dogs have four legs). In contrast, a property that has only a causal or statistical connection to the kind is not understood to be an aspect of being that kind of thing and does not support formal explanations of the property or generate normative expectations for the property (Prasada et al., 2013).

The formal structure of kind representations can thus be augmented to include a component that represents the properties to which the kind has principled connections. This component represents the kind's character and distinguishes it from other kinds. The revised kind mechanism is given here:

(7) K_i—< $a_1, a_2 \ldots, a_n$

In (7), the kind mechanism is understood as projecting (—<) an aspect structure ($a_1, a_2, \ldots a_n$) that represents the properties that a kind has by virtue of being that kind of thing. These properties are understood to be aspects of being that kind of thing and collectively characterize and distinguish that kind from other kinds. The augmented mechanism (7) incorporates a component in terms of which kinds are qualitatively distinguished from one another. Furthermore, the properties that characterize a kind and distinguish it from other kinds are lawfully connected to that kind and thus support formal explanation of the properties and the expectation that the property will generally be present in instances. Figure 4.2 shows the augmented mechanism that includes the component that represents the qualitative character of the kind. The character of the kind is represented via the properties to which the kind has principled connections. Those properties are represented as aspects of being that kind of thing ($a_1, a_2, a_3 \ldots$).

The core mechanism represented in figure 4.2 incorporates the four essential characteristics of how we think about kinds and their instances identified here.

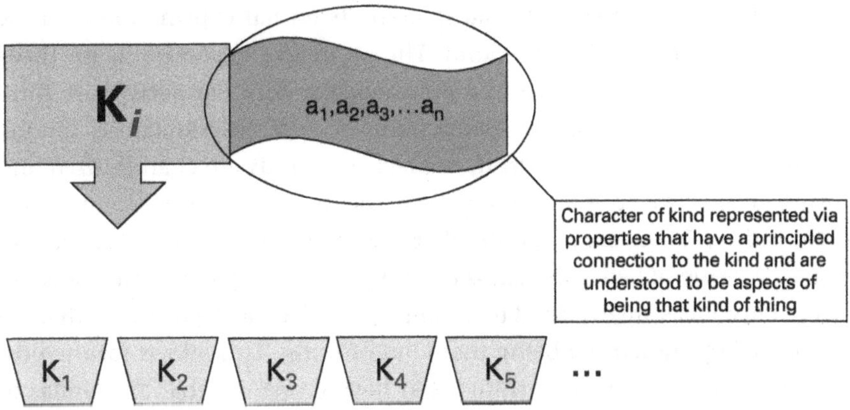

Figure 4.2
A mechanism for representing kinds updated to include a component that represents the character of the kind via the properties that have a principled connection to the kind. Those properties are represented as aspects of being that kind of thing. *Source:* Author.

4.2.3 The Proposed Representation of Kind Concepts Provides the Instance-of-Kind Perspective

Although the mechanisms for representing kinds and instances of kinds in figure 4.2 were not directly motivated by the instance-of-kind perspective, the mechanisms provide the instance-of-kind perspective. The instance-of-kind representations (6) generated by the kind mechanism (7) represent an instance as simultaneously unique and utterly nonunique within the kind. Each instance-of-kind representation has an index that is unique to the instance and a representation of the kind that the instance belongs to (which is shared by all instances of the kind). Furthermore, each instance-of-kind representation is implicitly contained in the kind representation (5) that generates the instance-of-kind representations, and as such, the mechanisms support a member-of relation between instance-of-kind representations and the kind representation. Because categorization is achieved by activating an instance-of-kind representation (6) rather than through the combination of an untyped particular and a type representation, the particulars that we think and talk about are intrinsically typed. Furthermore, because instances of a kind are expected to have the properties that characterize and distinguish the kinds of things they are, there is an explanatory link between the kind of thing that something is and some of the properties such that it supports formal explanation of those properties in instances of the kind. The normative expectations for those properties that are provided by the instance-of-kind perspective are similarly explained. Finally, the specification of expected similarities among instances of a kind follows from the fact that kinds are characterized and distinguished from one another by properties to which they have principled connections. These are properties that instances of the kind are expected to have by virtue of being the kinds of things that they are, and thus ways in which instances are expected to be similar. They are also properties that are explained by something being that kind of thing. This allows for the possibility that there could be things that have properties that are similar to instances of a given kind of thing, but whose properties are not explained by being that kind of thing, and thus those things are not understood to be instances of that kind despite their similarity to instances of that kind. Similarly, it allows for the possibility that there can be things with properties that are quite dissimilar from those of most instances, and yet be explained by being that kind of thing. Thus we see how the proposed mechanism can capture the expectation that instances of a kind will be similar without making similarity the basis for categorization. Furthermore, we see that the proposed mechanism captures the characteristics of the instance-of-kind perspective.

4.3 Elaborating the Core Mechanism for Representing Kind Concepts

4.3.1 Four Additional Characteristics of How We Think and Talk About Kinds and Instances of Kinds

The mechanism in figure 4.2 captures core characteristics of the ways in which we think and talk about kinds and their instances. Here, I identify four characteristics that allow us to elaborate the mechanism along two dimensions—to represent the mode of existence of instances and kinds and to represent the intrinsic hierarchical structure of kinds.

(v) Instances of kinds can exist actually or potentially. When we think of instances of a kind, the instances *can* be instances that actually exist, but they need not be. We are perfectly capable of thinking of other, potentially existing instances. Potentially existing instances of a kind differ numerically from actually existing instances of the kind—they are different instances of that kind. They also differ in their mode of existence—they exist potentially, not actually. As is the case with numerical difference, difference in mode of existence does not entail any qualitative differences. Thus, we can easily think of two qualitatively identical twin dogs, Fido and Spot, and wonder what it would have been like if they had had an identical triplet sibling, Rover. Potentially existing instances of a kind don't have to be qualitatively identical to actually existing instances, but they could be. Potentially and actually existing instances need not be distinguished qualitatively. Potentially existing instances of a kind must differ from actually existing instances of that kind numerically and in their mode of existence, but they need not differ qualitatively.

(vi) Kinds exist actually or not at all. Unlike instances of kinds, which can be thought to exist actually or potentially, kinds can only be thought of as existing actually. Whereas there are many kinds that do not actually exist but could exist, there aren't any potentially existing kinds that differ from actually existing kinds only in their mode of existence. The only mode of existence that a specific kind can have is actual existence. Kinds exist actually or not at all. Kinds that do not exist at a given point in time do not differ from kinds that actually exist only in their mode of existence; they must also differ in their qualitative character.[1] Instances of kinds, on the other hand, can exist actually as well as potentially.

(vii) Kinds can intrinsically contain instances and subkinds. Although in the first instance, we think of kinds (e.g., dogs) as containing instances of kinds (e.g., Fido, Rover), we can also think of kinds as containing subkinds (e.g., collies, dachshunds), which, in turn, contain instances (e.g., Fido, Lassie). These two ways of thinking about kinds is revealed by the fact that sentences such as (8) can be interpreted as quantifying over instances of kinds (9) or subkinds (10):

(8) The two elephants at the zoo differ in their diet preferences.

(9) Dumbo likes peanuts, but Babar does not.

(10) African elephants like peanuts, but Asian elephants do not.

In interpreting sentence (8) as talking about instances of kinds, we understand it to be a statement about two individual elephants such as Dumbo and Babar (9). When we understand sentence (8) as quantifying over kinds, on the other hand, the statement is about two kinds of elephants such as African elephants and Asian elephants (10), and thus it is not necessarily about two individual elephants, but rather is about the unlimited number of instances of the two kinds of elephants that have instances at the zoo.

(viii) Subkinds represent different ways of being that kind of thing. The kinds contained in a kind differ from one another in principled ways and thus specify different ways of being that kind of thing. For example, collies and dachshunds are different kinds of dogs, and as such, they differ from one another in principled ways and specify two different ways of being a dog. As such, they provide two different ways of realizing those properties that characterize dogs. Instances of kinds, on the other hand, do not differ in principled ways, and thus different instances are understood to simply instantiate the kind in question. They do not instantiate the kind in different ways. Fido and Spot are understood to be different dogs, but they are not understood to exemplify different ways of being a dog. This difference is reflected in the fact that subkinds are understood to be related to kinds via a *kind-of* relation, whereas instances are related to kinds via an *is-a* relation. As such, thinking about hierarchical relations concerning kinds implicates both class inclusion (*is a*) and kind specification (*kind of*) relations.

4.3.2 Elaborating the Formal Mechanisms to Capture These Characteristics of How We Think of Kinds and Instances of Kinds

Instances of a kind can differ in their mode of existence, with some instances of a kind existing actually and others existing only potentially. The instance-of kind representations in (6) and figures 4.1 and 4.2 do not mark the mode of existence of an instance. As such, they must be augmented to do so. I propose that we do this by adding an untyped index (I use the Greek letters α, β, γ . . . to represent untyped indices) to the instance-of kind representations that are used to think about actually existing instances. The untyped index, unlike the typed indices that distinguish instances within a kind, is not generated by the kind mechanism. Thus, in (11) and figure 4.3, K_1, K_2, and K_3 are instance-of-kind representations of actually existing instances of kind K, and K_4 and K_5 are instance-of-kind representations of potentially existing instances of kind K.

(11) $_{\alpha}K_1 \quad _{\beta}K_2 \quad _{\gamma}K_3 \quad K_4 \quad K_5 \ldots$

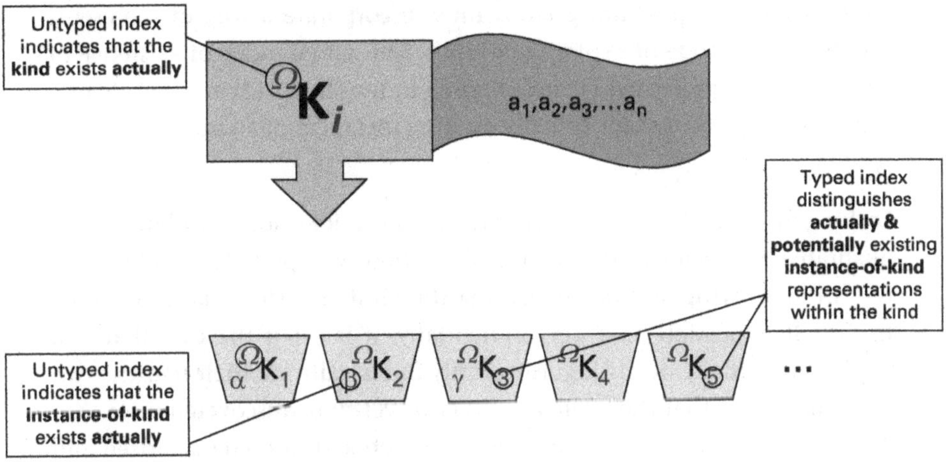

Figure 4.3
Mechanisms for representing kinds and instances of kinds, with elaborations to represent the mode of existence of kinds and instances of kinds. *Source:* Author.

Because actual existence is represented via an untyped index that has no intrinsic descriptive content, actually and potentially existing instances need not differ qualitatively.

Some evidence that our instance-of-kind representations have the formal structure depicted in figure 4.3 comes from Prasada and Hall (2019). They provided evidence that participants would not allow the transformation of an actually existing instance of a given kind into a different actually or potentially existing instance of the *same* kind, but they would allow the possibility that it could be turned into a potentially (but not actually) existing instance of a *different* kind. So, for example, participants did not think that one dog could be transformed into any other (actually or potentially existing) dog, but they did allow that it could be changed to a potentially (but not actually) existing cat. This pattern of constraints is predicted by the typed index that distinguishes each (actually and potentially existing) instance-of-kind representation within a kind and the untyped index that distinguishes an actually existing instance-of-kind representation from other actually existing instances of the same and different kinds, but not from potentially existing of either the same or different kind.

Unlike instances of kinds, which are understood to exist potentially as well as actually, kinds are understood to exist actually only. As such, the mechanism for representing kinds must incorporate a way of representing the actual mode of existence of kinds. The actual mode of existence of kinds can be represented via an untyped index (Ω in (12) and figure 4.3), just as it is represented for actually existing instances of a kind:[2]

(12) ${}^{\Omega}\mathrm{K}\vdash\!\!-\!\!< a_1, a_2 \ldots, a_n$

Lacking intrinsic descriptive content, the index does not contribute to the qualitative character of the kind. It does, however, provide an additional way of identifying and distinguishing the kind as a single kind of thing. In fact, it is possible that developmentally, it is often the case that two kinds may initially be distinguished only indexically via their names, with the expectation that they will also differ in yet-to-be-discovered ways (Xu, 2010). The yet-to-be-discovered ways in which a kind is characterized and distinguished from other kinds are specified by the component of (12) that represents the character of the kind. That character qualitatively characterizes and distinguishes the kind via the properties that have principled connections to the kind and are represented as aspects of being that kind of thing. Those properties are expected to generally be present in instances of the kind. Consistent with this, infants expect nouns to name categories characterized by shared properties from an early age (Waxman & Markow, 1995).

Kinds are understood to contain an unlimited number of instances, but they can also be thought of as containing subkinds. As such, the mechanism that represents kinds must be augmented (13) so it can generate subkind representations (15) in addition to instance-of-kind representations (14). The generative mechanism that represents kinds (13) thus implicitly contains both instances (represented via the variable *subscript* on the right side of the kind representation) and subkinds (represented via the variable *superscript* on the right side of the kind representation). All these elaborations are depicted in figure 4.4.

(13) ${}^{\Omega}\mathrm{K}^{s}\vdash\!\!-\!\!< a_1, a_2 \ldots, a_n$

(14) ${}_{\alpha}{}^{\Omega}\mathrm{K}_{1\beta}{}^{\Omega}\mathrm{K}_{2\,\gamma}{}^{\Omega}\mathrm{K}_{3}{}^{\Omega}\mathrm{K}_{4}{}^{\Omega}\mathrm{K}_{5}\ldots$

(15) ${}^{\delta}\mathrm{K}^{1}\vdash\!\!-\!\!< a^1{}_1, a^1{}_2 \ldots, a^1{}_{n,}{}^{\varepsilon}\mathrm{K}^{2}\vdash\!\!-\!\!< a^2{}_1, a^2{}_2 \ldots, a^2{}_{n,}{}^{\zeta}\mathrm{K}^{3}\vdash\!\!-\!\!< a^3{}_1, a^3{}_2 \ldots, a^3{}_n$

The subkind representations in (15) and figure 4.4 are distinguished from one another in three ways.

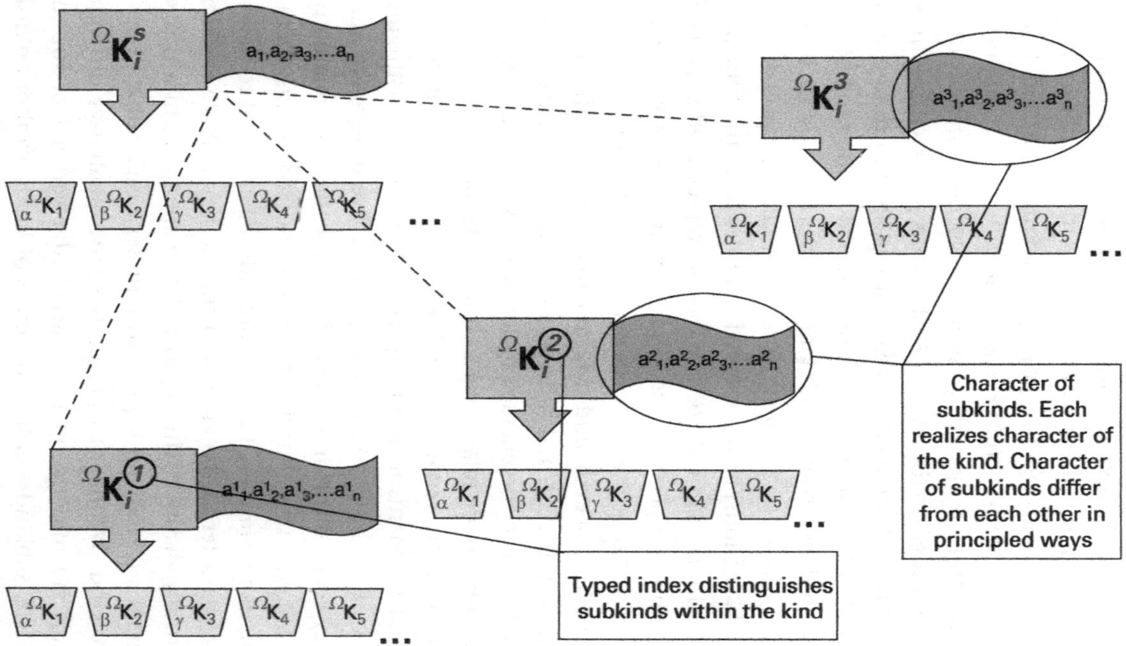

Figure 4.4

The mechanism for representing kinds, updated to capture the fact that kinds can be thought of as containing an unlimited number of instances as well as subkinds of the kind. *Source:* Author.

First, being kinds, subkinds exist actually, and thus each subkind representation has an untyped index (e.g., δ, ε, and ζ in (15)) to represent the subkind's actual mode of existence and distinguish it from other actually existing subkinds.

Second, being kinds, subkinds differ in qualitatively principled ways from one another, and thus the subkind representations differ in the properties to which each subkind has principled connections and thus the aspectual structure that they project (e.g.,—< $a^1_1, a^1_2 \ldots, a^1_n$). Importantly, the aspectual structures projected by subkinds do not differ in arbitrary ways from one another or the aspectual structure projected by the kind representation. Subkinds are qualitatively distinguished from one another because each subkind has a different way of realizing the properties that characterize the kind. For example, all kinds of dogs bark, but each kind of dog has its own bark—its own way of barking. As such, the properties that characterize a subkind must be represented at a finer grain than the properties that characterize the kind. Properties that are represented as accidental variants of the same property at the kind level are represented as being different ways of realizing the property when characterizing the subkind. Many accidental differences of kind-determined properties at the kind level are represented as nonaccidental differences at the subkind level. Consequently, subkinds realize the characteristics of the kind in different ways and kinds and subkinds are related via a kind-of relation.

Finally, subkind representations are distinguished by an index generated by the kind mechanism (right superscript on K) that distinguishes subkinds of a kind without assuming any qualitative differences among them. This may seem surprising, given that kinds must be qualitatively distinguished. But when we consider subkinds rather than kinds, there is a sense in which, like instances of kinds, subkinds are expected to have the properties that characterize the kind, and thus to be qualitatively identical but numerically distinct. The typed index helps to capture this way of thinking about subkinds, as well as to represent each subkind as one subkind of the kind. Within this perspective, the qualitative differences among subkinds are understood to be accidental variants of the same quality. Of course, subkinds, unlike instances of kinds, cannot be distinguished only numerically; they must also be distinguished qualitatively, which requires us to reconstrue the qualities at a finer grain in the manner described previously.

Given that the subkind representations are kind representations that generate instance-of-kind representations, we must consider how the instance-of-kind representations generated by the subkind representations are related to the instance-of-kind representations generated by the kind representation. Each instance-of-kind representation generated by the subkind is identified with an instance-of-kind representation of the kind, but not vice versa. This captures the intuition that every instance of a subkind (e.g., every collie and every dachshund) is an instance of the kind (e.g., is a dog), but not every instance of a kind (e.g., dog) is an instance of a specific subkind (e.g., is a collie, or is a dachshund). As such, there is a class inclusion relation between the instances of a subkind and the instances of the kind. Furthermore, the representations allow that there may be instance-of-kind representations that do not map onto *any* subkind instance-of-kind representations. For example, some dogs (e.g., mutts) are just dogs without being an instance of some subkind of dog.

Finally, it should be noted that the kind mechanism generates the instance-of-kind representations and the subkind representations if the kind contains subkinds. The subkinds represent the intrinsic possibility of there being principled variants of the kind—that there may be systematically different ways of being that kind of thing. The properties that the kind has by virtue of being the kind of thing that it is are the properties that instances of the kind are supposed to have by virtue of being the kind of thing that they are. They are also the properties that are realized in different ways by different subkinds. The subkind representations simply represent different ways of being that kind of thing. For example, being a collie and being a dachshund are two ways of being a dog. Subkind representations such as COLLIE or DACHSHUND must be distinguished from other systematic ways of dividing a kind, such as dividing DOGS into HERDING DOGS, GUARD DOGS, or THERAPY DOGS.

Even if such divisions yield nonarbitrary categories, the resulting categories are not understood as merely being different ways of being a dog. Instead, each subcategory has a unifying principle of its own. This is why the subcategory representation cannot be formally generated by the kind mechanism, as the subcategories are not merely representing one way of being that kind of thing but instead possess some further unifying principle. The instances of such subcategories stand in a class inclusion relation to the instances of the kind; however, the kind is not intrinsically divided into such categories, and thus the subcategory representations do not intrinsically stand in a kind-of relation to the kind representation. This is why class inclusion statements about the relation between the instances of the subcategory and the instances of the kind sound natural (16), but kind realization statements between the subcategory representation and the kind representation sound odd (17) (see Prasada et al., 2002, for related experimental evidence):

(16) Herding dogs are dogs.

(17) #Herding dogs are a kind of dog.

4.4 Additional Phenomena Explained by the Proposed Mechanism

Although the proposed mechanism for representing kind concepts was not designed to do so, it helps explain additional conceptual phenomena that are not accounted for by existing approaches, and, in some cases, are problematic for existing approaches. These include the following.

First, kinds are understood to have *indefinitely many* instances. This characteristic of kinds is puzzling if kinds are represented as sets, but it is easily captured by the proposed mechanisms. Sets can contain a finite number of elements (e.g., the set of natural numbers between 2 and 9) or an infinite number of elements (e.g., the set of natural numbers). In either case, the set has a definite cardinality. In contrast, we cannot simply ask about how many instances are in a kind. It sounds odd to ask, for example, "How many dogs are there?" We can ask instead: "How many dogs there are in this house now?" or "How many dogs were in this house yesterday evening?" But we would not ask how many dogs were there. This is because there can be different numbers of dogs at different times and places. The kind "dog" does not have a definite number of dogs in it. This characteristic of kinds is problematic for set representations because the identity of a set is given by the identity of its elements, and thus there cannot be different elements, much less different numbers of elements, in a set at different times. The proposed mechanisms easily capture this characteristic of kinds because the generative mechanism that represents the kind (13), and thus the kind's identity, is independent of how many instance-of-kind representations (14) exist at any given time (14). The proposed mechanisms thus allow for the possibility that a kind can have different numbers of instances of a kind at different times—that a kind has indefinitely many instances in it.

Second, we can characterize a kind in terms of a property that is not exemplified in any of the instances of that kind (Krifka et al., 1995). For example, "Orange-Crusher-2000s crush oranges" can be understood to be true even if no instance of the category ORANGE-CRUSHER-2000 has ever been used to crush an orange. This is puzzling if kinds are summary representations, as none of the instances has the characteristic that characterizes the kind. Similarly, such cases are problematic for thinking of kinds as sets, as none of the instances of the kind is an instance of the set of things that actually crush oranges. The proposed mechanisms have no problem capturing this fact about how we think and talk about kinds and their instances because the kind and the properties that characterize and distinguish it from other kinds are represented via the generative mechanism (13) that represents the kind via a principled connection between the kind and the property. As such, instances of the kind are supposed to have the property, but they need not.

Third, we can think and talk about a kind even though no instances of the kind actually exist. For example, a company may have developed plans for producing two kinds of artifacts, the *OC-2000,* which crushes oranges, and the *WP-3000,* which pulverizes watermelons, and then pondered the potential advantages of actually producing one or the other. In this scenario, we have no problems thinking about a kind that has no actual instances and distinguishing it from another kind that also has no actual instances. Such cases are problematic for set-theoretic approaches but pose no problems for the proposed theory because the generative mechanisms that represent and characterize and distinguish kinds (13) do not require that there are any actually existing instance-of-kind representations (14) that have been generated by the kind mechanism.

Fourth, we assume that an instance of a basic-level kind remains that kind of thing throughout its existence. What allows us to do this? On standard views of categorization, categorization is achieved by type representations being predicated of, or applied to, the representation of an untyped particular (e.g., Jackendoff, 1983; Barsalou, 2012). Given that the type representations are predicates, they are formally similar to other predicates, such as "hungry," "running," "puppy," "pet," or "nuisance," and the range of times and circumstances in which the predicate applies to the subject is not determined formally, but rather on the basis of our knowledge regarding the predicate and the subject. The problem, then, is how we come to know the temporal extent of each predicate, and in particular the temporal extent of the basic-level kind predicate.[3] The stability of kind identity has no formal basis within that approach. That contrasts sharply with the theory of kind representation being developed here, which provides a formal basis for the stability of kind identity while also providing a formal basis for the possibility of an entity changing kinds.

In contrast to the standard view, the current theory proposes that the things that we think and talk about are intrinsically typed and identified via instance-of-kind (14) representations. Activation of an instance-of-kind representation identifies what we are thinking about. As such, they can be the subjects of which we can predicate various things. For any predicable that is predicated of the instance-of-kind representation, one must specify the temporal extent of its applicability to the subject. Note, however, that the kind identity of the subject is not provided via predication. The kind identity is intrinsic to the subject representation (14), which has a component that represents the kind and a component that represents which instance of the kind the subject is. As neither component is independently represented, there is no question about the temporal extent over which a type representation applies to an untyped token representation. The form of the representation simply identifies what something is *atemporally*, which provides the basis for the intuition that kind identity is unchanging throughout a thing's existence. As such, the proposed theory provides an account of where the intuition of kind stability comes from whereas the standard approach fails to do so.

Fifth, the current theory also accounts for the possibility of kind identity changing in certain extraordinary circumstances. Recall that instance-of-kind representations can be of actually existing or potentially existing instances of the kind and the representations of actually existing instances have an additional untyped index. The index represents the fact that actually existing instances have an existence over and above their intelligible existence as instances of a kind. The untyped index provides a way of identifying the instance that is over and above its identification as an instance of a kind. As such, it provides the formal basis for how something could be understood to change kind identity. It represents that which is an instance of a kind, and that under extraordinary circumstances, and usually by mysterious means, may become an instance of some other kind. Although the index provides the means by which we can think of something as changing its kind identity, it does not identify or characterize what that something is. Thus, while even very young children can understand a story in which a fairy godmother changes a pumpkin into a carriage, neither the children, nor we adults, can say what it is that remained the same.[4] The untyped index captures the intuition that something persists, but also that we do not know what that is. Finally, because we do not know what remains the same during the change, only that *something* does, the means by which the transformation happens must also be mysterious (unintelligible) to us (e.g., magical and supernatural). Thus, the proposed mechanisms can account for the general stability of kind identity while also allowing for the possibility of a change in kind identity in extraordinary situations.[5]

Sixth, there are ways of talking about a kind that arise from the fact that individual members display a given characteristic (e.g., tires are black), as well as characteristics of a kind that involve the kind itself rather than being true of individual members of the kind (e.g., dogs are widespread; dogs evolved from wolves). To capture such differences, one needs to be able to represent the kind itself in a distinct manner from the representation of the instances that belong to the kind, as well as the relation between these two types of representations (Krifka et al., 1995; Chierchia, 1998; Ritchie & Knobe, 2020). The mechanisms in (13) and (14) do precisely that.

4.5 The Nature of Conceptual Knowledge Represented by the Conceptual Form of Kind Concepts

The structure of the mechanism depicted in figure 4.4 was motivated by eight virtually undeniable facts about how we think and talk about kinds, instances of kinds, and subkinds of kinds. That mechanism embodies those eight characteristics, but it also captures the characteristics of the instance-of-kind perspective discussed in chapter 3, despite the fact that the mechanism was not designed to do so. Furthermore, it also captures the six additional facts about how we think and talk about kinds and their instances discussed in section 4.4. The fact that the proposed theory can account for these additional nonobvious phenomena provides further evidence for the proposed conceptual form of kind concepts.

The conceptual form in figure 4.4 provides instructions for how to think about kinds, how they differ from one another, how instances of kinds and subkinds of kinds are expected to be similar to one another and to differ from one another, and how instances and subkinds are related to kinds and the properties that characterize and distinguish each kind. In so doing, the conceptual form captures the ways in which we can think about kinds and their instances and subkinds.

Although classes of kind concepts differ in terms of the components of conceptual form that are intrinsic to their representations (see chapter 5), kind representations in all classes contain the component that represents the character of the kind in terms of the properties that the kind has by virtue of being the kind of thing that it is. Those are properties that have principled connections to the kind and are understood to be aspects of being that kind of thing. Instances of the kind are supposed to have them, and the presence of the property in instances of the kind is explained by their being the kind of thing that they are. They are also expected to exist in instances of the kind generally. Forming a representation of a specific kind concept, such as DOG, involves specification of at least some of the properties to which the kind has principled connections.

Specification of the conceptual form of kind concepts results in the specification of a set of *principles* for thinking about that kind and its instances (e.g., "Dogs bark" or "Dogs have four legs").[6] The principles specify starting points for thinking about the kind and its instances—what we have reason to expect to be true simply on the basis that something is that kind of thing. As such, the conceptual form of a concept provides us a way of thinking and talking about instances of a kind about which we have only partial information, or even no specific information, about them.

Because the conceptual form of kind concepts provides *principles* for thinking and talking about things, the knowledge is *unrevisable* on the basis of empirical evidence. The principles do not describe what things are like, but instead specify the properties instances of kinds are supposed to have because they are the kinds of things they are. Consequently, if something fails to have a property that it is supposed to have because it is the kind of thing that it is, this does not count as evidence against the principle, as the principle does not describe how things are. In fact, the principle leads us to understand the lack of the relevant property in an instance of the kind as being due to some reason *other* than the thing being what it is. As such, the principle, unlike a mere fact or generalization, leads us to explain away putative counterexamples, further insulating it from revision on the basis of empirical evidence.[7] For this reason, the knowledge encoded in the conceptual form of kind concepts is *stable*. Because the knowledge provided by the formal dimension of concepts is stable, it is also *potentially linguistically relevant*. For example, in English, indefinite singular generics (e.g., "A dog has four legs") are licensed only for properties that have a principled connection to the kind (Prasada & Dillingham, 2009). Other examples of the relevance of conceptual form to linguistic expression are found in table 1.1 and discussed in chapters 6, 7, and 8.

The knowledge embodied in the conceptual form of kind concepts *complements* other forms of conceptual knowledge. The conceptual form of kind concepts specifies the properties that are understood to be aspects of being that kind of thing and that instances are supposed to have simply because they are the kinds of things that they are. However, kinds can also be characterized in terms of properties that are not understood to be aspects of being that kind of thing, but that are causally connected to the material natures of the kind (e.g., "Mosquitoes carry West Nile virus"). Furthermore, kinds can be characterized in terms of properties that are statistically prevalent in instances of a kind, even though they are neither understood to be an aspect of being that kind of thing nor causally connected to the thing's material nature (Prasada et al., 2013). In such cases, there are presumably stable environmental causes or social structures (Vasilyeva & Lombrozo, 2020) that are responsible for the nonaccidental connection between the kind and the property. These ways of characterizing kinds complement the ways in which the kind is characterized via its conceptual form. It is also possible that in many cases, these ways of characterizing kinds are not mutually exclusive and kind-characterizing properties may be understood as being related to the kind in multiple ways.

4.6 An Addition to the Conceptual Form of Kind Concepts

The conceptual form in figure 4.4 depicts the formal structure of lexically expressible kind concepts such as DOG and TREE. That structure provides instructions for how to think about such kinds and their instances and subkinds. As such, it captures the way in which those kinds, instances, and subkinds are known to us. For all kinds, we understand certain properties to characterize it because it is the kind of thing that it is. Those properties are understood to be aspects of being that kind of thing and support formal explanations that explain the properties by reference to the kind of thing that something is (e.g., "Birds have wings because they are birds"). In many cases, those properties can also support teleological explanations that explain by referring to the sake for which the kind has the property (e.g., "Birds have wings for the sake of flying"). Furthermore, the properties can also support causal explanations (e.g., "There is something about birds that causes them to have wings"). In such cases, the conceptual form of the kind identifies the properties to be explained, and thus even though the teleological and causal explanations may complement and flesh out our understanding of the kind, those understandings depend on the way that things are known to us via the principled connections that characterize the kind.

Natural kinds such as "dogs," "trees," "water," and "wood" are understood to have an existence that is independent of our conceptions of them. Conceptual representations for such kinds must make room for understandings of the kinds as they are in and of themselves, in addition to how such kinds are known to us via the understandings embodied in their conceptual form. In general, we do not know what an understanding of dogs/trees/water/wood . . . as they are in themselves might be, but we do recognize the possibility of developing such an understanding via scientific inquiry. This contrasts with our understanding of artifact kinds like "tables," "forks," and "cars." Those kinds are what we understand them to be. We do not expect that scientific inquiry will reveal to us what those kinds are in themselves, independent of our conceptions of them. As such, we need to add a placeholder to the structure of kind concepts such as DOG, TREE, WATER and WOOD to indicate the possibility of an understanding of the kind as it is in itself, which is independent of the understanding of the kind as it is known to us via the principled connections that characterize the kind. The modified structure with the place holder for concepts like DOG and TREE is given in figure 4.5.

The conceptual form in figure 4.5 captures the possibility of developing an understanding of the kind that is independent of the conception of the kind provided by the conceptual form. It also allows for the possibility of scientific understandings coexisting with understandings that are rooted in our commonsense conceptions of things (Goldberg & Thomson-Schill, 2009; Shtulman & Valcarcel, 2012; Kelemen, Rottman, & Seston, 2013; Legare & Shtulman, 2018; Shtulman & Young, 2024). As we will see in chapter 6, it also provides a representational basis for the essence assumption that is part of psychological essentialism.

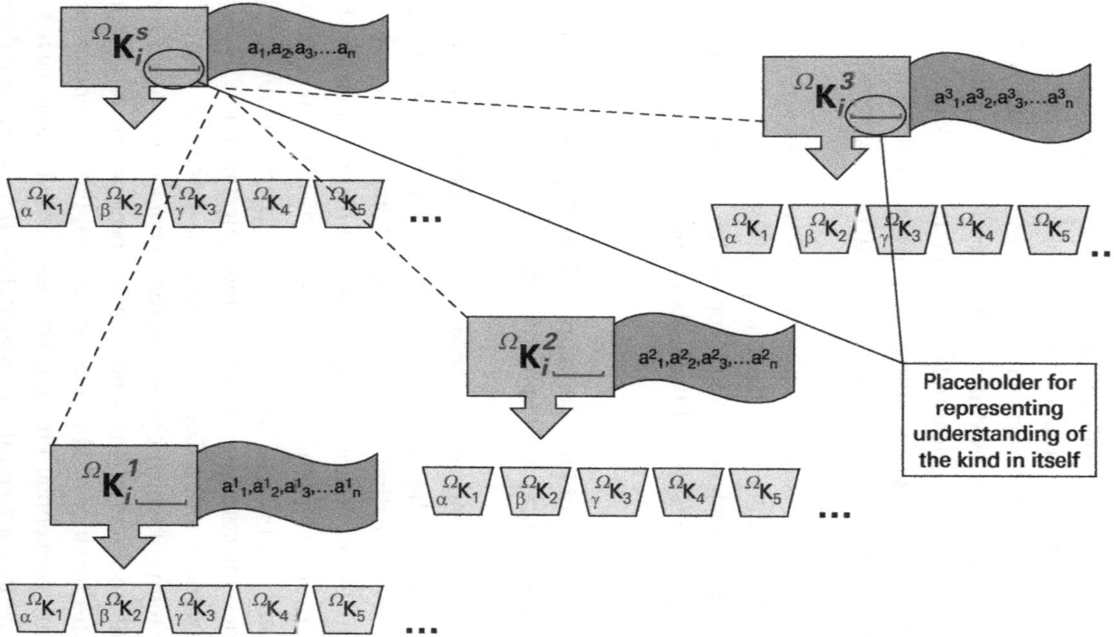

Figure 4.5

The formal structure of kind concepts such as DOG, TREE, WOOD, and SAND, which contain placeholders for representing our understanding of kinds as they are in themselves, independent of the conception of the kinds as they are known to us via the understanding embodied in the conceptual form of the kind concept. *Source:* Author.

Another advantage of the conceptual form depicted in figure 4.5 is that it identifies the dimensions along which the proposed mechanisms can be elaborated to capture the constrained variety of perspectives provided by different classes of kind concepts. Chapter 5 shows how we can generate the perspectives provided by a wide variety of classes of kind concepts via various combinations of the components of the mechanism depicted in figure 4.5 and formally distinct ways of realizing those components. Chapters 6, 7, and 8 provide further evidence of the proposed conceptual form of kind concepts by illustrating how the conceptual form of kind concepts also helps explain phenomena concerning the systematic polysemy of kind terms, the interpretation of metaphors, and a range of facts about generic knowledge and generic language.

Before elaborating the theory and showing how it helps explain those phenomena, I briefly consider how the conceptual form of kind concepts accounts for key characteristics of existing approaches to conceptual representation, as well as how we should think about the atomicity of conceptual form.

4.7 The Conceptual Form of Kind Concepts and the Key Characteristics of Existing Theories of Conceptual Representation

The conceptual form of kind concepts displays key characteristics of the way that existing approaches represent conceptual knowledge. In line with the theory- or explanation-based approach to conceptual representation, the conceptual form of kind concepts does not merely categorize but also specifies a way of understanding why instances of a kind have certain properties. Furthermore, because kind representations are presupposed by framework theories, the conceptual form of kind concepts can be incorporated into framework theories. Consistent with the prototype approach, the properties in terms of which conceptual form characterizes kinds need not be true of all instances of the kind. Finally, because the kind representation generates instance-of-kind representations, the conceptual form theory can capture conceptual phenomena that exemplar approaches to conceptual representation account for. As such, the conceptual form theory is well suited for capturing many of the phenomena studied by existing approaches to conceptual representation.

4.8 A Note on the Structure of Kind Concepts

The conceptual form of kind concepts has a component that represents the character of the kind via the properties that have principled connections to the kind, each of which is understood to represent an aspect of being that kind of thing. For example, the conceptual form of CANARY represents being yellow as an aspect of being a canary. Given that, one might wonder whether CANARY is a complex concept that has the concept YELLOW as one of its constituents?

It is not. To see why, it is helpful to compare CANARY to the concept YELLOW CANARY, which is uncontroversially a complex concept that has the concept YELLOW as one of its constituents. Every instance of YELLOW CANARY must be yellow, but instances of CANARY need not be yellow (e.g., albino canaries or canaries that have been painted red). Furthermore, one could *not* have the concept YELLOW CANARY without having the concept YELLOW, but one *could* have the concept CANARY without having the concept YELLOW, or while having the concept YELLOW but not knowing that being yellow is one aspect of being a canary. All of this shows that we should not think of YELLOW as a constituent of CANARY.

The aspect representations within the conceptual form of kind concepts are like the argument positions projected by verbs. The existence and identity of the aspect representations is determined by the class of kind concepts to which the kind concept belongs (e.g., the class of animal kind concepts), but the value of an aspect representation must be determined by looking at instances of the specific kind in question. The aspect representations do not contain conceptual information, but they identify the type of conceptual information that is understood to be an aspect of being that kind of thing. Specifying what the color of canaries is requires coindexing the aspect representation for the color dimension with the concept YELLOW (see Prasada & Dillingham, 2009). So, for example, the knowledge that canaries are yellow is represented in (18) by coindexing the concept YELLOW with one of the aspects (a_1) of the concept CANARY. The formula in (18) can be read as representing the thought that being yellow is one aspect of being a canary:

(18) CANARYi–< a_1, a_2, . . . YELLOWa_1

We are thus able to represent the fact that being yellow is an aspect of being a canary without making YELLOW a constituent of CANARY. Lexically expressible kind concepts like CANARY are thus atomic, in that they lack constituent structure (Fodor, 1998a). Lexically expressible kind concepts like CANARY *are atomic, but not isolated.* That is because representing a kind concept requires specifying at least some of the properties that the kind has by virtue of being the kind of thing that it is, and thus coindexing at least some of its aspect representations with concepts that represent the value to which the kind has a principled connection. The atomic structure of lexically expressible kind concepts of a given class affords establishing principled connections between the kind and certain types of concepts that represent aspects of being that kind of thing.

5 Generating Perspectives

Concepts for kinds such as DOG, CAT, and TREE provide perspectives for thinking and talking about things as instances of kinds as well as kinds and subkinds, as such. As important as such basic-level object kind concepts are to our cognitive lives (Rosch, 1978), they are only a small fraction of lexically expressible kind concepts, and thus we must determine the conceptual forms of other kind concepts and how they may be generated.

One possibility is that all kind concepts have the same conceptual form, and thus the conceptual form of the class of kind concepts investigated in chapter 4 is the conceptual form found in all kind concepts. We can call this the *Uniform Conceptual Form* hypothesis. According to this hypothesis, all kind concepts have the same conceptual form, and thus kind concepts are distinguished from one another only by their content—by the specific qualitative characteristics to which the kinds have principled connections. Just as the concepts DOG and CAT are distinguished by the properties that they have by virtue of being the kinds of things that they are, concepts such as TABLE, CHAIR, ANIMAL, FURNITURE, PET, and POEM are also distinguished from one another and from concepts such as DOG, CAT, and TREE only on the basis of the properties they have principled connections to. On this view, all kind concepts provide the same perspectives for thinking and talking about things, but they differ in the properties the kinds have by virtue of their being the kinds of things they are. Existing approaches to conceptual representation share with the Uniform Conceptual Form hypothesis the idea that concepts differ only in their content. Within standard approaches, the form of a conceptual representation is assumed to be that of a definition, a prototype, or a theory, with concepts differing on the features that are part of the definition, prototype, or theory.[1] If the Uniform Conceptual Form hypothesis were correct, there would be no need to figure out how to generate the conceptual form of kind concepts other than the ones considered in chapter 4 because the conceptual form is the same for all kind concepts.

In contrast, the *Constrained Variety of Conceptual Forms* hypothesis holds that not all kind concepts have the same conceptual form. Kind concepts can be distinguished from one another by their conceptual forms as well as their content. Classes of kind concepts that are formally distinguished from one another provide different perspectives from which to think and talk about kinds and their instances. Kind concepts *within* a class are distinguished from one another by the properties to which they have principled connections, but not by the perspective that they provide for thinking and talking about those kinds and their properties. As will become abundantly clear, the evidence strongly favors the Constrained Variety of Conceptual Forms hypothesis over the Uniform Conceptual Form hypothesis. As such, we need a way to generate the variety of conceptual forms found in different classes of kind concepts.

In this chapter, I show how the perspectives provided by a wide range of lexically expressible kind concepts can be generated via various combinations of the components of the conceptual form of the class of kind concepts discussed in chapter 4, and by elaborating the components of the conceptual form of those kind concepts.

5.1 Kinds That Intrinsically Contain Instances and Subkinds

As discussed in chapter 4, concepts such as DOG, CAT, and TREE have the conceptual form shown in figure 4.5 (repeated here as figure 5.1). That conceptual form allows us to think about a kind and its instances, as well as subkinds of the kind, and it tends to be used to think and talk about natural object kinds. The conceptual form specifies how instance-of-kind representations are distinguished from one another, as well as how subkind-of-kind representations are distinguished from one another (if they exist). Furthermore, each instance is understood to be quantitatively equivalent, as is each subkind. As such, the conceptual form of kind concepts in this class specifies both instance-of-kind and subkind-of kind as units of quantification. Furthermore, both the kinds and subkinds in this class of concepts contain placeholders for representing an understanding of the kind that is independent of the understanding provided in terms of the properties that have a principled connection to the kind.

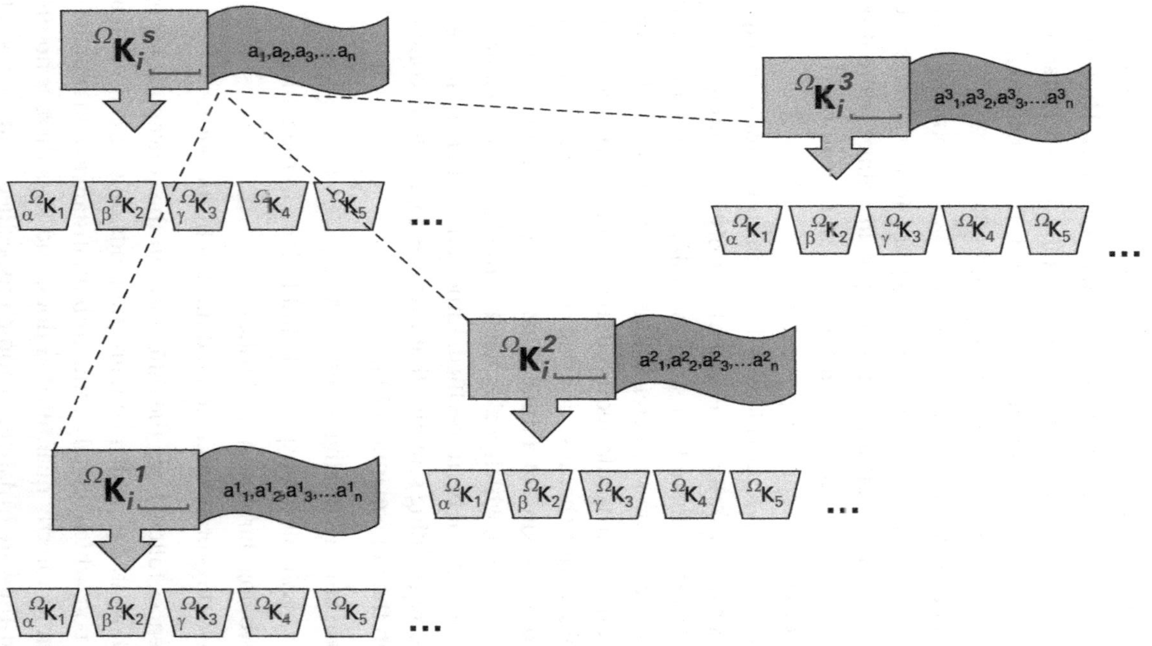

Figure 5.1

The conceptual form of the class of kind concepts that intrinsically contain instances and subkinds. *Source*: Author.

5.2 Kinds That Intrinsically Contain Subkinds, but No Instances (Directly)

Many basic-level kind concepts are understood as containing both instances and subkinds (e.g., RACCOON, DOG, FLOWER); however, there are other concepts (e.g., ANIMAL, VEHICLE) that only directly contain subkinds. These superordinate kind concepts contain instances only indirectly—as instances of the subkinds that are directly contained in the kind. The conceptual form of a superordinate concept such as ANIMAL is given in figure 5.2.

Some evidence that the conceptual form of ANIMAL contains kinds only directly comes from the fact that the conceptual form captures the intuition that there can be nothing that is simply an animal without being a specific kind of animal. This basic intuition is not accounted for by standard approaches to conceptual representation.

Given the perspective provided by the conceptual form of superordinate concepts such as ANIMAL, one cannot identify something as simply an instance of a superordinate kind such as "animal" (ANIMAL does not generate any instance-of-kind representations). If one were to identify something as an animal, it would require leaving out information about the instance's kind identity because nothing could be an animal without being an instance of some kind of animal. This is why (19) is interpreted in the manner given in (19′), which makes explicit the fact that the kind identity of the instances has not been specified. Importantly, there is no pressure to interpret a corresponding sentence that involves a basic-level concept such as DOG (20) in the same manner (20′):

(19) I saw two animals lurking in the bushes.

(19′) I saw two animals of unspecified kind lurking in the bushes.

(20) I saw two dogs lurking in the bushes.

(20′) I saw two dogs of unspecified kind lurking in the bushes.

These differences follow from the differences in the conceptual forms of a concept such as ANIMAL, which contains subkinds only directly, and a concept such as DOG, which directly contains both instances and subkinds. The instances can be directly quantified as dogs, and thus there is no pressure to think that they are instances of some unspecified kind of dog. On the other hand, there are no instance-of-kind representations generated by ANIMAL, and thus any instances must be instances of some unspecified kinds of animal.

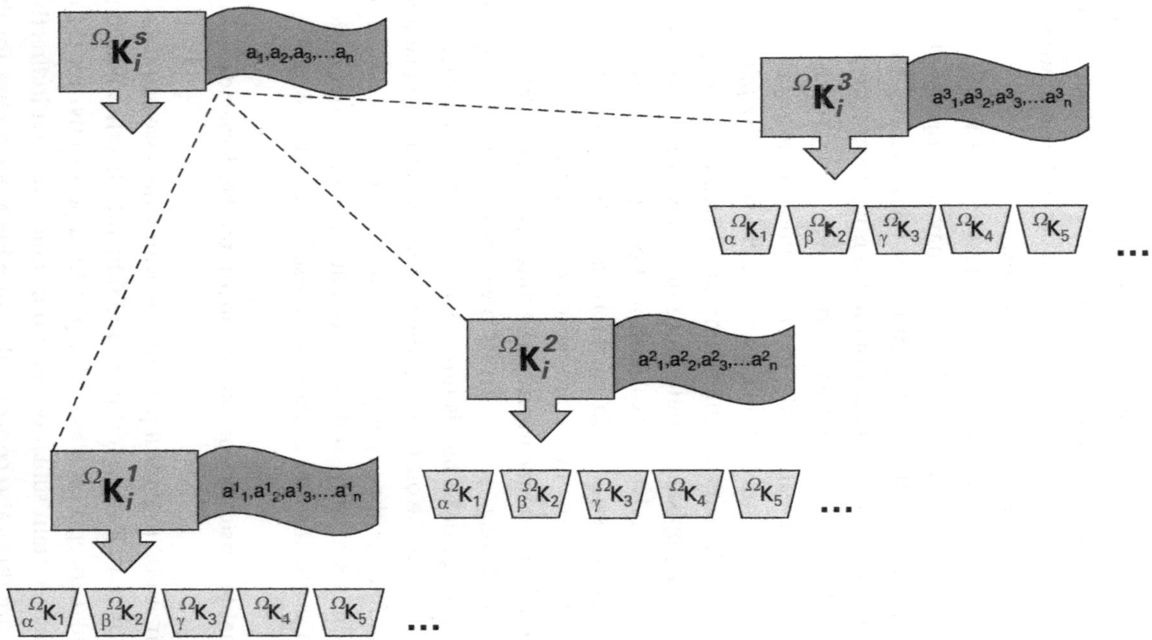

Figure 5.2
Conceptual form of kinds that intrinsically contain subkinds, but no instances (at least not directly). *Source:* Author.

The conceptual form of ANIMAL provides a perspective for thinking and talking about instances in which the kind identity of the instances is left unspecified. Under what conditions would such a perspective make sense? If the instances that are being quantified do not share a kind identity, their shared identity at the superordinate level allows them to be quantified. This provides a semantic basis for the tendency to assume that the two animals referred to in (19) are different kinds of animals.

There can also be pragmatic reasons for leaving the kind identity of the instances unspecified. The kind identity may be unspecified because the speaker is ignorant of the kind identities of the instances, the speaker believes that the kind identity of the instances is not relevant, or the speaker is motivated to obfuscate or be underinformative.

These observations reveal how the indirect representation of instances in the conceptual form of concepts such as ANIMAL can specify semantic conditions on the use of the superordinate concept, as well as trigger pragmatic inferences about the epistemic state and intentions of the speaker. None of these phenomena receive an explanation if concepts such as ANIMAL are understood to have the same conceptual form as concepts such as DOG, with the concepts differing only in the properties to which they have principled connections. It is hard to imagine how the difference in the properties that characterize animals (e.g., moves on its own, can perceive things, reproduces) and dogs (e.g., has a tail, is furry, barks) even begins to account for such phenomena. On the other hand, they receive a natural explanation if the conceptual form of the superordinate concepts such as ANIMAL directly contains subkinds and only indirectly contains instances, but basic-level concepts such as DOG directly contain both instances and subkinds.

5.3 Kinds That Intrinsically Contain Instances, but No Subkinds

Concepts such as DOG provide a perspective from which we can think and talk about kinds, as well as the instances and subkinds that they contain. On the other hand, concepts such as ANIMAL provide a perspective for thinking about kinds that contain subkinds, but no instances (except indirectly, via its subkinds). Are there also concepts that provide a perspective for thinking of kinds as intrinsically containing instances but not subkinds? The concept ROOK provides an example of such a concept. A rook is a chess piece that can move any number of spaces vertically or horizontally. Although

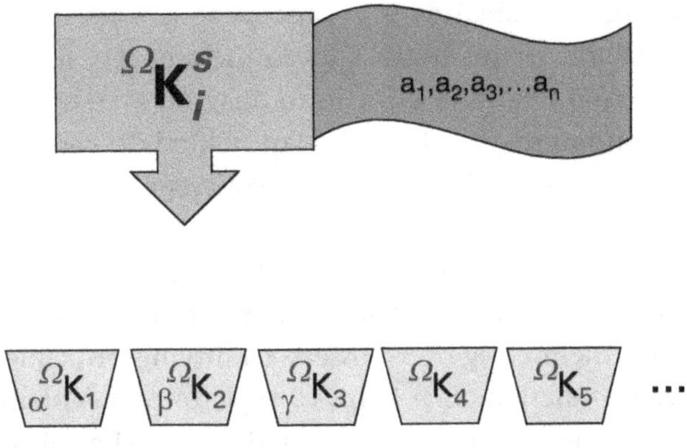

Figure 5.3
Conceptual form of kinds that intrinsically contain instances, but no subkinds. *Source:* Author.

there are many ways in which individual rooks may differ, in their shape, material constitution, color, size, and other characteristics, given the fact that there is an arbitrary connection between the form and function of rooks, it does not appear that there can be principled differences between rooks such that they would be thought to be different kinds—that there could be systematically different ways of being a rook. As such, concepts such as ROOK provide a perspective from which to think and talk about kinds and their instances in a manner that does not also allow us to think of the kind as containing subkinds. The variation among instances of such kinds can only be thought to be accidental—it cannot be understood to be principled, as that would suggest that there are subkinds within the kind.

It is worth noting that the same is true of the lowest-level kind in a kind hierarchy. For example, we probably think that the variation among collies is only accidental, as there are no principled differences between collies that would lead us to think that there are different kinds of collies. The difference between COLLIE and ROOK is that the lack of principled variation among collies is a potentially revisable fact about them, whereas the perspective provided by ROOK is one in which not only is there no principled variation among its instances, but there cannot be intrinsic principled variation in being that kind of thing.

I have used the kind concept ROOK to illustrate the conceptual form of a class of kind concepts that intrinsically contains instances, but does not intrinsically contain subkinds. The concept ROOK is ideal for illustrating this class of kind concepts because given the arbitrary relation between form and function, it is very difficult to imagine what a principled subkind of this concept may be. But this class of kind concepts also includes concepts such as TABLE, which *does* have coherent subcategories such as COFFEE TABLE, PICNIC TABLE, and DINING TABLE. Although these are systematic and coherent subcategories of table, they are not intrinsic subkinds of TABLE. They are not understood as simply different ways of being a table. Instead, these categories possess a unifying principle pertaining to their specific functions that helps to explain their existence and why their properties cluster in the way that they do. The subcategories do not merely represent different ways of being that kind of thing (e.g., a table). As such, the formal structure of the subcategory representations cannot be formally generated from the kind representation, as is the case for the classes of kind concepts that can intrinsically contain subkinds. It is for this reason that although we can choose to think and talk about principled subsets of concepts like TABLE as subkinds, we need not do so (Ritchie & Prasada, 2024). For example, suppose that someone asks, "What is a nightstand?" We could choose to provide an answer in which we identify NIGHTSTAND as a subkind (e.g., "A nightstand is a kind of table that . . ."), but we need not do so. We could equally well say something like, "A nightstand is an artifact that people make to. . . . ," which does not identify the category as a subkind of TABLE. Note how this contrasts with concepts like COLLIE, where an adequate answer requires identifying it as a subkind of DOG. Simply identifying collies as animals that have characteristics x,y, and z is not satisfactory. This difference between classes that have intrinsic subkinds and those that do not is reflected in the forms of systematic polysemy displayed by names for kinds from the two classes. For example, "dog" can be used to talk about the kind ("Dogs evolved from wolves"), its instances ("Two dogs I like are Jett and Fly") and its subkinds ("Two dogs I like are collies and dachshunds"), but "table" can be used to talk about the kind ("Tables were invented by the Egyptians") and its instances ("Two tables I bought are in the next room"), but not its subkinds (#"That store sells two tables—picnic tables and coffee tables") unless they are explicitly identified as such ("That store sells two *kinds of* tables—picnic tables and coffee tables"). For an explanation of these and related phenomena, see chapter 7 and Ritchie and Prasada (2024).

Finally, the kinds in this class of concepts do not contain a placeholder for representing an understanding of the kind that is independent of the understandings provided in terms of the properties that have a principled connection to the kind.

5.4 Kinds That Individuate Subkinds but Whose Instances Are Ad Hoc

Basic-level object kind concepts such as DOG provide a perspective for thinking about instances of the kind as well as subkinds. This explains why statements such as (21) are ambiguous and can be understood as referring to two individual dogs or two kinds of dogs. Concepts such as SAND are like the class of kind concepts that contains DOG, in that they provide the means for thinking and talking about subkinds. They differ, however, in that concepts like SAND cannot, on their own, be used to think or talk about instances of the kind. Thus (22) can be used to talk about two kinds of sand but not two instances of sand. To think and talk about two instances of sand, we must combine the kind concept with a concept such as PILE, MOUND, or HEAP to contextually identify instances of that kind and specify the perspective from which to think and talk about the instances (e.g., as piles/mounds/heaps/etc. . . .) (23):

(21) John likes only two dogs.

(22) There are two sands on this island.

(23) There are two piles of sand on this island.

Instance-of-kind representations for kinds like SAND are generated on an ad hoc basis in conjunction with an individuating concept. As such, the ad hoc instance-of-kind representations have an untyped index that represents the instance's actuality and distinguishes it from other actually existing ad hoc instances of that and other kinds. The conceptual form of such concepts is shown in figure 5.4.

The instance-of-kind representations for kind concepts like SAND lack a typed index that represents the instance as one of an unlimited number of potentially existing instances within the kind and distinguishes it from all instances of the kind. Evidence for that comes from Prasada and Hall (2019), who found that although participants do not allow one instance of a kind like DOG to be transformed into a different potentially existing instance of that same kind, they do allow such a possibility for kind concepts like SAND, which do not specify their instances and for which instances can be arbitrarily specified.

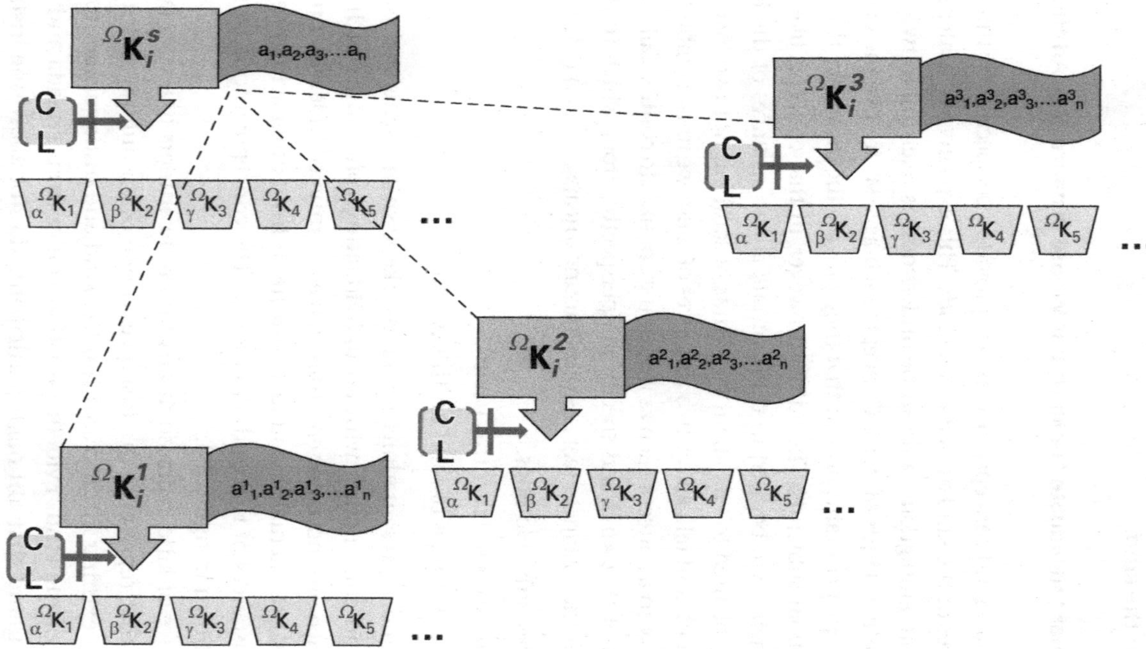

Figure 5.4
Conceptual form of kinds that intrinsically contain subkinds and ad hoc instances. The ad hoc instances are generated by combining the kind concept with an individuating concept. *Source:* Author.

In sum, kind concepts such as SAND provide the means for thinking about the kind sand and subkinds of sand, and, in combination with contextually specified individuating concepts such as PORTION, PILE and HEAP, they allow us to think and talk about instances of those kinds. Importantly, because the instance-of-kind representations are generated via conceptual combination, they can be generated using different individuating concepts on different occasions, and thus we cannot assume quantitative equivalence for ad hoc instances of a kind. As such, ad hoc instance-of-kind representations are both syntactically and semantically/conceptually different from instance-of-kind representations that are intrinsically generated by kind representations.

Finally, both the kinds and subkinds in this class of concepts contain a placeholder for representing an understanding of the kind that is independent of the understanding provided in terms of the properties that have a principled connection to the kind.

5.5 Kinds That Intrinsically Contain Both Abstract and Concrete Instances

Canonical basic-level object kind concepts such as DOG are understood as containing abstract subkinds (e.g., collie, dachshund) that are qualitatively distinguished from one another in a principled manner, as well as concrete instances (e.g., Fido, Spot) that are distinguished numerically with qualitative differences between instances understood as being accidental. However, basic-level object kind concepts like BOOK and NECKLACE are understood as intrinsically containing *both* abstract and concrete instances, but not subkinds.[2] Thus, (24) refers to an abstract instance of the kind "book," whereas (25) refers to a concrete instance of the kind "book":[3]

(24) My favorite book is *Midnight's Children*.

(25) My favorite book has a coffee stain on it.

(26) Sally and Susan bought the same necklace.

(27) Sally bought her necklace at the mall whereas Susan got hers via mail.

Similarly, (26) refers to an abstract instance of the kind "necklace," whereas (27) refers to concrete instances of the kind "necklace."

Concrete instances of kinds need differ from one another only numerically. Thus it is possible for all the properties that are not determined by the kind to be the same in every instance of the kind, and thus that all the instances of the kind are qualitatively identical. While this is possible, it is unlikely and not expected. Unless there is reason to think otherwise, we assume that the differences between instances of a kind are accidental. As such, the set of properties of an instance of a kind that do not characterize the kind—the *instance properties*—are assumed to vary arbitrarily from instance to instance (within the range determined by the kind). The only unity that the instance properties of a given instance are assumed to possess is an accidental unity. Whereas this is the case for the instances of many concepts, such as DOG and PAPER CLIP, it is not the case for instances of concepts like NECKLACE and BOOK. The properties of instances of such kinds possess a nonaccidental unity and coherence that is responsible for the aesthetic character and intelligibility of the instances. For example, the colors, shapes, textures, patterns, and arrangements of those elements give a necklace its aesthetic character. This aesthetic character can be shared by any number of concrete instances of the kind "necklace," and it characterizes and distinguishes an abstract instance of the kind from other abstract instances of that kind. Similarly, the sentences, paragraphs, chapters, illustrations, and arrangements of those elements constitute the intelligible content that can be realized in any number of concrete instances of the kind "book." That intelligible content characterizes and distinguishes the abstract instances of BOOK.

The abstract instances of a kind bear a one-many relation to the concrete instances of that kind. Any abstract instance may be realized in an any number of concrete instances of that kind. Each concrete instance possesses the abstract (aesthetic/intelligible) character of only one abstract instance of that kind.[4] In standing in a one-many relation to concrete instances of the kind, abstract instances are like kinds. However, they differ from kinds in that the abstract instances are intrinsically singular, whereas kinds are intrinsically plural. The abstract instances may be realized in any number of concrete instances; however, those concrete instances are not contained in the abstract instance. In contrast, the unlimited number of concrete instances of a kind are contained in the kind. This is why the kind must be represented via a generative mechanism, but the abstract instances may be represented via a representation similar to instance-of-kind representations.

Abstract-instance representations differ from concrete instance-of-kind representations in two ways. First, like kind representations, they contain

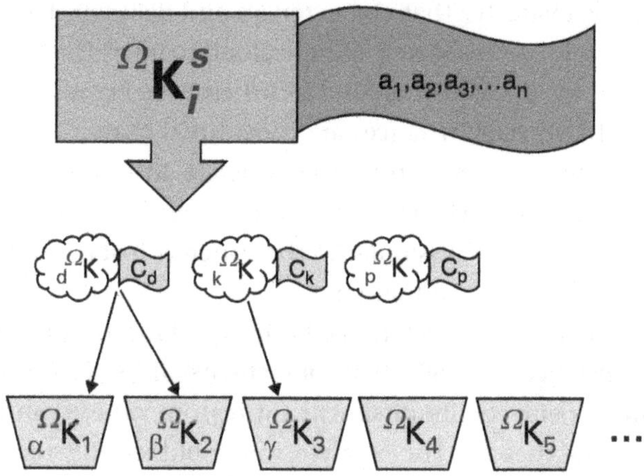

Figure 5.5
Conceptual form of kinds that intrinsically contain both abstract and concrete instances. *Source:* Author.

a component that represents the qualitative (aesthetic/intelligible) character of the instance that distinguishes it from other abstract instances. Second, the abstract-instance representations do not have a typed index, as we do not think of the various abstract instances of a kind as differing from one another only numerically and in their mode of existence. Abstract instances of a kind necessarily differ from one another in their aesthetic/intelligible character, and thus they cannot differ only numerically. For the same reason, putative potentially existing abstract instances cannot differ from actually existing abstract instances only in their mode of existence. Thus, *Shmyntactic Structures* cannot differ from *Syntactic Structures* except that it exists potentially, while the latter exists actually.

Because kind concepts like BOOK and NECKLACE are understood to contain both abstract and concrete instances, each of which are distinguished from one another in different ways, it is possible to quantify things like books and necklaces in two different ways. This is why (28) can have two correct answers:

(28) How many books/necklaces does Iris have?

Although abstract-instance representations and concrete instance-of-kind representations are distinct and provide different ways of quantifying instances of the kind, they are systematically related to one another, albeit in different ways depending on the type of abstract instance.

The aesthetic character that characterizes and distinguishes an abstract instance of a kind like "necklace" is understood to be realized in the matter of the concrete instances of the kind. Furthermore, because the aesthetic characters of the abstract instances are constituted of sensible matter (e.g., colors, shapes, textures, patterns), the concrete instances corresponding to a given abstract instance are expected to be perceptually similar to one another and perceptually dissimilar to the concrete instances that realize a different abstract instance of that kind.

These are the default expectations. Both expectations can be violated in specific circumstances, though. Two concrete instances of the same abstract instance can be perceptually dissimilar for reasons other than their being the necklaces that they are. For example, one could sensibly say (29), which attributes the perceptual dissimilarity to external causes:

(29) Sally and Sue bought the same necklace, but their necklaces do not look alike because someone painted Sally's necklace and put stickers on it.

The expectation that the concrete instances of different abstract instances are perceptually dissimilar can be violated because aesthetic characters are individuated by their sensory characteristics, as well as by the identity of the creator of the aesthetic character. Thus, it makes sense to say things like (30). These considerations suggest that the component of the conceptual form that distinguishes abstract instances must represent both the types of elements that the aesthetic character is constituted of (i.e., sensory matter) and the identity of the creator of the aesthetic character. We will consider this and other elaborations of the conceptual form of kind representations in section 5.7.

(30) Sally and Sue bought necklaces that look virtually identical, but they did not buy the same necklace. Sally's necklace was designed by Bob Smith, and Sue's necklace was designed by Bob Jones.

The intelligibility of the abstract instances of kind concepts such as BOOK derives from being constituted of intelligible matter. Consequently, we do not expect concrete instances of the same abstract instance to necessarily be perceptually similar. Instead, we expect concrete instances of the same abstract instance to be *understood* in the same manner and instances of different abstract instances to be understood differently.

The abstract instances, which are actually existing instances, are also understood to be potentially existing constituents of concrete instances of the kind, which are understood to be constituted of both intelligible and concrete matter. If the abstract instance were not represented as potentially a constituent of concrete instances, the abstract instance would not be a book, but merely a story or some information. This points to the need to elaborate the conceptual forms of kind representations and instance-of-kind representations to include the types of elements in terms of which specific types of kinds (e.g., dogs/necklaces/books) and their instances are constituted and characterized. We turn to that task (and other elaborations) in section 5.7.

Finally, the kinds in this class of concepts, like other artifact kinds, do not contain a placeholder for representing an understanding of the kind that is independent of the understanding provided in terms of the properties that have a principled connection to the kind.

5.6 Kinds That Are Characterized and Individuated in Both Abstract and Concrete Terms

The conceptual form of kind concepts such as BOOK provides a perspective in which instances of the kind can be characterized and individuated in both concrete and abstract terms. There are also kind concepts such as SCIENTIST, which provide both concrete and abstract ways of characterizing and individuating the kind. Knobe, Prasada, and Newman (2013) present evidence that concepts such as SCIENTIST allow us to categorize instances in terms of both abstract values and concrete features that realize those abstract values. For example, scientists can be characterized in terms of a set of descriptive features such as "collect data," "run experiments," "perform statistics," "publish in journals," etc. . . . , or in abstract terms such as "They seek to understand the principles that underlie phenomena by systematically observing, experimenting, and reasoning about phenomena." These two ways of characterizing a kind are not generally independent of one another. The concrete descriptive features are understood to have principled connections to the kind, and thus to be true of instances by virtue of being the kinds of things that they are. They are also understood as realizing the abstract values that characterize the kind. In addition to characterizing the kinds, the abstract values are understood as *explaining* why instances of that kind are characterized in terms of those concrete descriptive features. The concrete features provide the means by which the values that characterize the kind are realized. Consequently, we generally expect instances of the category to display both the abstract values and concrete characteristics of the kind.

Because properties that have a principled connection to a kind can fail to hold for instances of that kind, it is possible for instances of the kind to have the abstract values that characterize the kind, but not the concrete features that canonically realize those values. For example, someone who has not been trained as a scientist, nor works professionally as a scientist, but carefully observes, analyzes, and seeks to understand why things are the way they are in making everyday decisions may be thought to be a true scientist.

Furthermore, because it is possible to explain the concrete features of an instance of a kind via reasons other than the values that characterize the kind, there can be instances of the category that possess the concrete features that characterize the kind for reasons other than their being the means for realizing the abstract values that characterize the kind. An example of this would be someone who was trained as a scientist, works professionally as a scientist, publishes in scientific journals, and does other scientific activities but has no interest in discovering how things in the world really work and is only interested in fame and fortune and is willing to make up data. Such a person may be thought of as being a scientist in some sense but is not a true scientist.

Thus, while the two ways of characterizing concepts such as SCIENTIST are not independent, they nevertheless allow for the possibility of creating circumstances in which only one or the other way of characterizing the kind is possible and thus to provide distinct bases for categorization in such circumstances—a possibility exploited by the experiments reported in Knobe et al. (2013).

Dual character concepts such as SCIENTIST suggest that the component of kind representations that characterizes and distinguishes kinds from one another needs to be elaborated so it can contain concrete descriptive characteristics as well as abstract values and an explanatory link between the two whereby the abstract values are represented as explaining the concrete features.[5,6]

5.7 Classes of Kinds That Are Distinguished by Their Formal Explanatory Structure

The conceptual forms of all classes of kind concepts provide a perspective in which certain properties of instances of a kind are rendered intelligible via formal explanation. These are properties that have a principled connection to the kind and are understood to be an aspect of being that kind of thing.

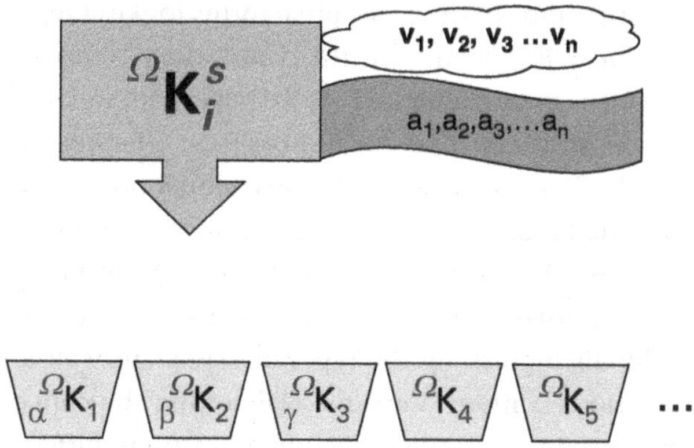

Figure 5.6
Conceptual form of kinds that are characterized in both abstract and concrete terms.
Source: Author.

In its current form, the representation provides a perspective in which we expect there will be some properties that instances have because they are the kind of thing that they are, but the conceptual form provides no guidance as to what those properties may be for any given kind.

Furthermore, the conceptual form provides no explanation for the observation that the number and types of properties that have principled connections depend on the type of kind (Prasada & Dillingham, 2006). For example, kinds of living things generally have more principled connections than do artifact kinds; while there are principled connections between kinds of living things and the stuff of which they are constituted, this is not the case for artifact kinds. Artifact kinds constrain but do not determine the kinds of stuff of which they are constituted (Prasada, 2017).

The conceptual forms of kind concepts that we have considered so far provide a perspective in which all kinds have the same explanatory structure. However, it has been understood since at least the time of Aristotle that explanatory structure varies from domain to domain, and thus distinct modes of explanation are available for understanding things and their characteristics across domains (e.g., Moravcsik, 1975, 1981, 1998; Keil, 1994; Pustejovsky, 1995). Insofar as the proposed conceptual form provides no basis for how the explanatory structure of different types of kinds may differ across domains, it fails to capture an important difference in the perspectives that we take when thinking and talking about kinds of things from distinct domains.

Given these limitations of the conceptual forms for kind representations considered thus far, it is evident that the explanatory component of kind representations needs to be elaborated. As we have seen, we can distinguish different classes of kind concepts on the basis on whether they intrinsically individuate both instances and subkinds or only one or the other, and also on whether the kinds and instances are individuated in both abstract and concrete terms. However, the component of conceptual form that represents the qualitative character of a kind currently does not provide a basis for formally distinguishing the types of explanatory structure found in kinds from various content domains. Elaborating the conceptual form of object kind concepts to formally distinguish among different types of properties would make it possible to capture domain differences via the number and types of properties that kinds have principled connections to.[7] It would also help capture differences in the types of explanatory structure supported by kind concepts in different domains.

How then should we distinguish the property types that kinds have principled connections to? Given the intrinsic connection between kind representations and explanation, types of kinds should be distinguished by having principled connections to different types of explanatorily significant properties (Prasada, 2017). Aristotle identified four distinct forms of explanation: (1) Explanations that make reference to what something is (formal explanations); (2) explanations that make reference to what something is constituted of (material explanations); (3) explanations that make reference to the end or sake-for-which something is or is the way it is (teleological explanations); and (4) explanations that make reference to that which generates the thing to be explained (efficient explanations). The conceptual forms of different types of kinds can be generated by representing principled connections to the explanatorily significant properties for one or more of these modes of explanation. According to this proposal, kind concepts in different domains have principled connections to different numbers and types of explanatorily significant properties. Further distinctions among types of kinds are generated by the various ways in which the explanatorily significant properties for each mode of explanation may be realized. Figures 5.7a and 5.7b illustrate how different types of kinds can be generated by representing principled connections between a kind and the explanatorily significant properties for different modes of explanation and the ways in which those types of properties may be realized.

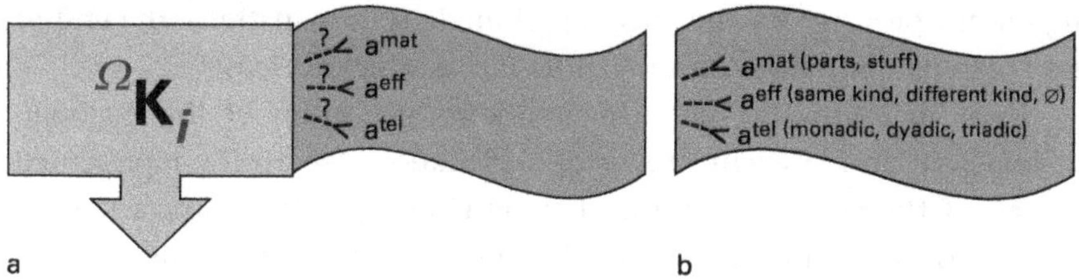

Figure 5.7

(a) Schema for generating kind representations that have formally distinct characters. Formally distinct characters result from kind representations having a principled connection to one or more of the types of properties that are explanatorily significant in the material (*mat*), efficient (*eff*), and teleological (*tel*) modes of explanation. (b) The properties that are explanatorily significant in the material, efficient, and teleological modes of explanation may be realized in different ways. The *mat* properties may be realized via the parts and/or the stuff of which instances of the kind are constituted. The efficient component can represent instances of the kind as being produced by the same kind of thing (living things), a different kind of thing (artifacts), or no specific kind of thing (natural objects like mountains, which are produced by any appropriate combination of causal forces). The teleological component can be realized via monadic (e.g., growth), dyadic (e.g., goal-directed movement), and triadic (e.g., reason) functional capacities. *Source:* Author.

It is worth noting that the conceptual form of concepts like BOOK and SCIENTIST provide independent motivations for including the explanatorily significant properties for the material, teleological, and efficient modes of explanation in the conceptual form of kind representations. Specifically, as discussed previously, concepts like BOOK require representing what instances of the kind are constituted of, as well as the identity of the author as abstract instances of books are distinguished by both the intelligible matter that they are constituted of and the creator of the abstract instance. Furthermore, the concrete instances of such concepts are understood to be constituted, in part, by the abstract instances. As such, concepts like BOOK require that we represent the explanatorily significant property for the material mode of explanation—what instances of a kind are constituted of—and the explanatorily significant property for the efficient mode of explanation—that which generates an instance. Finally, concepts like SCIENTIST require representing the abstract values (ends) that characterize the kinds and explain and unify the concrete properties that characterize the kind. As such, concepts like SCIENTIST require that we represent the explanatorily significant property for the teleological mode of explanation—the end or sake for which instances or some of their characteristics exist.

Given that the explanatorily significant properties of the material, teleological, and efficient modes of explanation need to be represented as part of the conceptual form of some kind representations, a natural way of distinguishing types of kinds is by the number and types of such properties a given type of kind has principled connections to. Generating distinct types of kinds on the basis of the numbers and types of explanatorily significant properties that they have principled connections to and the manner in which they are realized results in types of kinds that correspond closely to distinctions that are ontologically significant from the perspective of commonsense conception: for example, living/nonliving, natural objects/artifacts/living things, and plants/animals/humans (see Prasada, 2017).[8]

Elaborating the conceptual form of kind representations to include principled connections to explanatorily significant property types means that the component that represents the qualitative character of the kind needs to be replaced by the numbers and types of explanatorily significant properties that they have principled connections to and the manner in which they are realized for different classes of kinds. Figure 5.8 illustrates how the conceptual form of concepts for animate kinds and artifact kinds differ. The fragment of conceptual form in the top part shows the types of properties that animate kind concepts such as SPIDER have principled connection to, and the bottom part shows the types of properties that artifact kind concepts such as CHAIR have principled connections to.

The set of conceptual forms that result from distinguishing types of kinds by the number and types of explanatorily significant properties that a kind has principled connections to and the manner in which those properties are realized capture important characteristics of how we think and talk about things from different content domains.[9] First, they capture Prasada and Dillingham's (2006) observation that the number and types of principled connections that a kind has appear to vary by domain. Second, the conceptual forms for each type of kind embody expectations about the types of properties that instances of kinds in a given domain are expected to have. Finally, the conceptual forms provide the basis for expecting distinct modes of explanation to be differentially applicable across domains. Because the distinct types of kinds specify principled connections to properties that

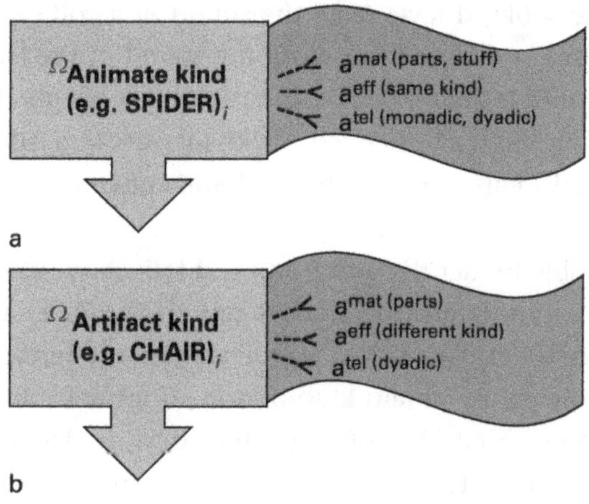

Figure 5.8
Kind representations from different classes are distinguished by the number and types of properties that they have principled connections to, as well as the ways in which they may be realized. This drawing illustrates how the formal structure of (a) animate kind representations (e.g., SPIDER) differs from that of (b) artifact kind representations (e.g., CHAIR). *Source:* Author.

are explanatorily significant for other modes of explanation (e.g., material, teleological, efficient), the formal explanatory structure of the kind concept provides the basis for representing these other modes of explanation. A large body of research shows that causal and teleological modes of explanation are available to preschoolers (e.g., Carey, 1985; Gelman, 1990; Keil, 1994; Kelemen, 1999; Sobel & Kirkham, 2006), in addition to the formal mode of explanation provided by the conceptual form of kind representations (Haward, Wagner, Carey, & Prasada, 2018).

5.8 Ad Hoc Subkinds

Both object kind concepts like DOG and substance kind concepts like SAND provide perspectives from which we can think and talk about subkinds. Each subkind is understood to realize the properties that characterize the kind in its own way. As such, establishing subkinds requires representing properties at a finer grain so that what is considered to be accidental variation at the kind level (e.g., the variability in the sound of barks of dogs) can be understood to be the result of a number of ways of realizing those properties at the subkind level (e.g., the sound of a collie's bark is understood to differ in a principled way from the sound of the bark of a dachshund and every other kind of dog). As such, the conceptual form of kind representations provides a principled way to identify and think about kinds and their subkinds. Furthermore, the kind and subkinds have similar conceptual forms.

It is also possible to identify and think and talk about subkinds in an ad hoc manner based on contextual considerations (Krifka et al., 1995; Mendia, 2019; Umbach & Gust, 2014).[10] For example, a statement such as (31) can be used to identify a certain kind of dog—dogs that bite—and make a generalization about that kind of dog—that they are dangerous. That the expression "dogs that bite" can be used to think and talk about a subkind rather than simply a collection of individual dogs that bite is evident from the fact that predicates such as "widespread," which apply to kinds, but not individual instances of kinds, can be felicitously predicated of it (32). Furthermore, "dogs that bite" can also be understood as quantifying over kinds (33):

(31) Dogs that bite are dangerous.

(32) Dogs that bite are widespread.

(33) Three dogs that bite are collies, dachshunds, and poodles.

The ad hoc subkind "dogs that bite" can consist of instances of many kinds of dog (e.g., some collies, some dachshunds, some poodles) and thus need not be similar to one another to a greater extent than dogs generally (unlike taxonomic subkinds). Note further that identifying an ad hoc subkind does not require representing properties at a finer grain of representation. Ad hoc subkinds can be established by combining a kind concept with other concepts that specify the *additional* characteristics (at the same grain of representation) that instances of the specified kind *must* have in order to be part of the ad hoc subkind (see figure 5.9).

As is evident in figure 5.9, the conceptual form of an ad hoc subkind is not that of a generative mechanism that directly generates instance-of-kind representations of the ad hoc subkind. Instead, the instance-of-kind representations that correspond to an ad hoc subkind are the instance-of kind representations of the kind that represents the head noun in the expression that is used to think about the ad hoc subkind that meet the additional conditions that are required for belonging to the ad hoc subkind. Thus,

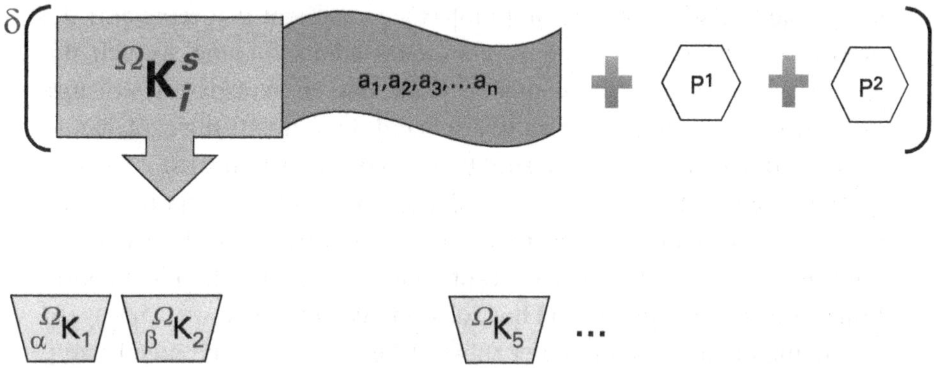

Figure 5.9
Conceptual form of ad hoc subkinds. *Source*: Author.

there are no unique instance-of-kind representations with their own typed index for ad hoc kinds. For example, there is no DOG THAT BITES$_{13}$. Instances of the ad hoc subkind are distinguished from one another in the same way that instances of the kind from which the ad hoc category are derived. Furthermore, the actually and potentially existing instances of the ad hoc category are nothing other than the actually and potentially existing instances of the kind from which the ad hoc kind is derived that meet the additional conditions specified by the ad hoc kind.

The conceptual form of ad hoc subkinds, being a representation of a kind, contains an untyped index that represents the fact that it exists actually. This captures the intuition that the kind is something other than the multiplicity of instances that belong to the kind (see Ritchie & Knobe, 2020).

Contextually relevant concerns specify properties that *must* be true of members of the ad hoc subkind by combining the kind concept, which provides the condition that the members of the ad hoc subkind must be instances of the kind, with the concepts that represent the *additional conditions* on membership in the ad hoc subkind. As there are no constraints on which additional conditions may be specified for membership in an ad hoc subkind, the ad hoc subkind cannot be assumed to possess a principled unity. This means that while we can make statistical or causal generalizations about the ad hoc subkind, the ad hoc subkind is not characterized and distinguished from other subkinds via principled connections to a set of properties. This is why, unlike canonical kind representations, the conceptual form of ad hoc subkinds does not project a component that represents the properties with a principled connection to the ad hoc subkind. As such, the ad hoc subkind is not represented as possessing an explanatory structure of its own. Nevertheless, the ad hoc subkind does inherit the explanatory structure that characterizes the kind that enters into defining the subkind.

The conceptual form of ad hoc subkinds contains two components that together specify conditions of application of the concept: (1) the generative kind mechanism and (2) the concepts that specify the additional conditions for membership in the ad hoc subkind. As such, there are two ways of generating subkinds of the ad hoc subkind. First, one can limit membership to subkinds of the kind that enter into specification of the ad hoc concept. For example, DACHSHUNDS THAT BITE is a subkind of the ad hoc subkind DOGS THAT BITE. Second, one can limit membership of a subkind of an ad hoc subkind by specifying additional concepts that impose additional conditions on membership in the subkind of the ad hoc subkind. For example, DOGS THAT BITE AND SCRATCH is a subkind of the ad hoc subkind DOGS THAT BITE. As such, there are principled as well as contextually determined ways of generating subkinds of ad hoc subkinds.

Canonical kind and subkind representations allow us to think about kinds and their instances in terms of the characteristics that they have by virtue of being the kinds of things that they are. Ad hoc subkinds allow us to construct contextually relevant perspectives for thinking about kinds and their instances. For example, DOGS THAT BITE provides a perspective for thinking about dogs in which dogs are identified by whether they bite or not, and thus the concept provides a starting point for discovering which properties may characterize and distinguish such dogs from ones that don't bite, and perhaps even discover the factors that are responsible for the cluster of characteristics displayed by members of the ad hoc subkind.[11]

5.9 Ad Hoc Kinds

Kind concepts such as SAND combine with contextually specified concepts to specify ad hoc *instances* of the kind. Kind concepts can also be combined with contextually specified concepts to identify ad hoc *subkinds* of kinds (e.g., DOGS THAT BITE AND SCRATCH). It turns out that it is also possible to represent an ad hoc kind by contextually specifying the conditions that must be satisfied in order to belong to the ad hoc kind.

Recall that in the canonical case, kinds contain both kinds and instances. As such, it should be possible to specify ad hoc kinds whose members include kinds and instances of kinds. For example, the ad hoc kind THINGS THAT BITE AND SCRATCH may be thought to include cats, raccoons, dogs, and several other kinds of things. It may also include Fluffy the cat, Rhonda the raccoon, Fido and Rover the dogs, and many other instances of one kind or another.

Ad hoc kinds (e.g., THINGS THAT BITE AND SCRATCH) are not represented via a generative mechanism. They are represented via a combination of concepts that specify a set of conditions (a description) that must be satisfied by the kinds and instances of kinds that are contained in the ad hoc kind. As such, ad hoc kind concepts *presuppose* kind representations and instance-of-kind representations.

Furthermore, ad hoc kind representations do not generate a typed index that represents instances as one of an unlimited number of actually or potentially existing instances. The members of ad hoc kinds are actually existing kinds and actually existing instances of kinds that satisfy the conditions of membership specified by the description that represents the ad hoc kind. There are no potentially existing instances of ad hoc kinds per se. Kinds that are contained in the ad hoc kind have potentially existing instances, and some of those instances may satisfy the conditions specified for membership in the ad hoc kind, so there can be an unlimited number of instances in the ad hoc kind even though there is no representation of potentially existing instances of the ad hoc kind. Similarly, the fact that there are no potentially existing instances of an ad hoc kind per se does not mean that there cannot be ad hoc kinds that do not have any instances. That will happen when the kinds whose actually existing instances would be instances of the ad hoc kind if they met the right conditions have no instances that meet those conditions.

It should be noted that not all descriptions allow us to represent an ad hoc kind. For example, the description THE THINGS THAT ARE ON MY DESK NOW includes a limited number of instances of some kinds (e.g., 1 computer, 2 books, 3 pads, 1 pen), but kinds have an unlimited number of instances in them, which is why they must be represented via a generative mechanism. Only descriptions that can characterize kinds (e.g., THINGS THAT BITE AND SCRATCH), which are represented via generative mechanisms, may represent ad hoc kinds.

Finally, ad hoc kinds cannot be assumed to possess a principled unity and thus do not project a component that represents the nonexistent properties that have a principled connection to the ad hoc kind. As such, the ad hoc kind is not represented as possessing any explanatory structure of its own.

5.10 Aspects of the Generative Theory of Kind Perspectives

As the foregoing makes abundantly clear, not all kind concepts have the same conceptual form, differing only in the properties that characterize each kind. Instead, kind concepts can differ from one another in both their conceptual form and the properties that characterize a kind. All of this is problematic for the Uniform Conceptual Form hypothesis, an analog of which appears to be implicit in most existing approaches to conceptual representation (see note 1).

In addition to showing that kind concepts display a large variety of conceptual forms, the foregoing illustrates how a wide range of kind concepts, including ad hoc kind concepts, can be generated from the components of the conceptual form of object kind concepts and some elaborations of those components.

Canonical kind concepts such as DOG are represented via a generative mechanism that has three key characteristics. First, it generates (and thus distinguishes) an unlimited number of instance-of-kind representations and thus captures the principled unity of the unlimited number of instances of a kind. Second, it possesses a character that specifies and unifies the properties that characterize and individuate the kind. In doing so, it specifies the properties that are shared by instances of the kind—properties they are supposed to have by virtue of being the kind of thing that they are. Third, it provides a principled basis for distinguishing subkinds of the kind—each subkind realizes the properties that the kind has by virtue of its being the kind it is in a different way.

As we saw, contextual information can sometimes be used to generate and distinguish ad hoc instances of a kind (e.g., piles/heaps/cups of sand); identify ad hoc subkinds on the basis of contextually specified characteristics for dividing a kind (e.g., DOGS THAT BITE); and even identify ad hoc kinds by contextually specifying the characteristics that must be shared by members of the kind (e.g., THINGS THAT BITE AND SCRATCH). Ad hoc instances, ad hoc subkinds, and ad hoc kinds are similar in some respects to instances, subkinds, and kinds represented via a generative kind mechanism. However, as noted previously, there are important ways in which they differ because the former are not generated in a principled manner.

Identifying ad hoc instances, ad hoc subkinds, and ad hoc kinds requires combining concepts and thus syntax to specify how to think about the relations between the elements combined. Given that syntax provides the means for generating perspectives in which elements are not intrinsically related to one another, relying on contextual information yields perspectives in which elements need not bear principled relations to one another. As such, ad hoc instances are understood to not be intrinsically related to the kinds of which they are instances; Ad hoc subkinds are understood not to involve intrinsic divisions of a kind; and ad hoc kinds contain instances and kinds that are not intrinsically related to one another.

By showing how the perspectives provided by a wide range of kind concepts, including ad hoc kind concepts, can be generated from the components of the conceptual form of natural object kind concepts and some elaborations of those components, this chapter has provided a sketch of a generative theory of kind representations. As discussed earlier, a generative theory of kind representation helps ameliorate questions regarding the acquisition, evolution, and linguistic expression of the perspectives provided by lexically expressible kind concepts.

Another important characteristic of the generative theory of conceptual forms sketched out here is that it does not use mechanisms for representing bare or completely unspecified kinds (Prasada, 2017, 2021). Instead, we have mechanisms for representing various types or classes of kinds, each of which has principled connections to different numbers and types of properties. The particular things that we think about must be thought about via the perspectives provided by the kind concepts of one or another class of kind concepts. So, for example, if the perspectives provided by the class of kind concepts that DOG belongs to are selected for thinking about something, we will think of the thing as one of an unlimited number of instances of that kind; we will also potentially be able to think of it as an instance of a subkind of the kind where that subkind is understood to simply be one way of being that kind of thing; think that the thing was produced by other instances of that kind (i.e., reproduced); think that it has the parts that it does because it is the kind of thing that it is; and that it has means for intrinsically dyadic functional capacities such as sensing and goal-directed movement. That is not a bare kind representation! Instead, the class of the kind concept makes available many abstract characteristics of the kind, and applying that formal structure to a particular thing allows us to specify at least how some of those abstract characteristics are realized in the thing being thought about.

The theory of how kind perspectives are generated that has been sketched out so far specifies the components in terms of which a wide variety of kind perspectives are structured. Future work must investigate how these components must be further elaborated or augmented to capture an even larger variety of perspectives that are provided by lexically expressible kind concepts. It must also investigate the dependencies that exist between distinct components and the constraints on how the components may or may not be combined to capture the perspectives provided by linguistically expressible kind concepts.

III Linguistic Applications

6 New Perspectives on Generics

Dogs bark, cars have wheels, barns are red, ducks lay eggs, and mosquitoes carry West Nile virus. Although we only ever encounter instances of kinds, never kinds as such, much of our world knowledge consists of generic knowledge—knowledge about kinds of things. Furthermore, children as young as two-and-a-half years old (and probably even younger) understand and use generic language to talk about kinds of things (Gelman & Tardif, 1998; Gelman & Raman, 2003; Gelman, 2003; Gelman, Goetz, Sarnecka, & Flukes, 2008; Graham, Gelman, & Clarke, 2016). In fact, deaf children whose hearing parents had not exposed the children to sign language develop gesture systems that contain generic utterances, suggesting that a language model is not necessary for young children to produce generic utterances. Those children, like hearing children exposed to linguistic input, also produced more generic utterances for animal kinds than artifact kinds, suggesting that this asymmetry reflects the different conceptual organizations underlying animal and artifact categories rather than linguistic input (Goldin-Meadow, Gelman, & Mylander, 2005). Although we certainly know many facts about specific instances of kinds, such as the family dog, the neighbor's dog, and our car, as well as facts about sets of instances of various kinds, such as the children in Ms. Thompson's fifth-grade class, the trees in our backyard, and the plates in the dishwasher, Cimpian (2016) reviews evidence suggesting that knowledge about categories or kinds of things is privileged in a number of ways over knowledge of instances of kinds or sets of instances of kinds. Children find it particularly easy to reason about categories, are particularly motivated to acquire information about categories, have particularly accurate information about categories, and assume that information about categories is widely known (Cimpian,

2016). These findings, as well as other data, suggest that generic knowledge and generic language are early emerging and fundamental aspects of human cognition (Gelman, 2003; Leslie, 2007, 2008).

Nevertheless, generic knowledge and generic language pose difficult challenges for those interested in figuring out how generic knowledge is acquired, how the meanings of generic sentences are represented, and how children acquire the complexities of generic language. My goal in this chapter is to show how the theory of conceptual form provides a new way of thinking about generics and many of the puzzles that they present, and that it even provides potential solutions to some of those puzzles. Given this goal, my discussion of generic phenomena and approaches to them is highly selective. For overviews of issues pertaining to generic knowledge and generic language, readers should consult Carlson and Pelletier (1995); Krifka et al. (1995); Cohen (2002); Mari, Beyssade, and Del Prete (2012); Sterken (2017); and Leslie and Lerner (2022). I will be focused on highlighting what I see as the advantages provided by the theory of conceptual form. Hopefully, what I say here will provide the impetus for future work to consider how the solutions proffered by the theory of conceptual form relate to those provided by other approaches in greater detail. Finally, it is important to note that I will be discussing only generic language phenomena in English; the reader should keep that in mind, as I will not repeatedly qualify my claims as such. Although many of the phenomena and explanations that I consider extend to similar phenomena in other languages, I leave it to future work to determine the extent of the language specificity of the claims that I make here.

6.1 Some Puzzling Phenomena Concerning Generics

6.1.1 Puzzling Proportions

Generics express general facts. Each of the sentences that follow—that is, (34)–(39)—tells us something that is generally true of instances of the kind referred to by the subject of the sentence rather than something that is true of some particular instance or set of instances of that kind. It is natural, then, to think that the predicates in these sentences must apply to all, virtually all, or at least the majority of instances of the relevant kinds. This intuition is borne out in cases like (34)–(36), as all triangles have three sides; and while not all dogs have four legs, virtually all of them do; and while it is certainly true that there are many barns that are not red, most barns are.

(34) Triangles have three sides.

(35) Dogs have four legs.

(36) Barns are red.

(37) Ducks lay eggs.

(38) Mosquitoes carry West Nile virus.

(39) #Human beings are right-handed.

However, as has often been noted, there are cases in which the predicate is true of a *minority* of the instances. For example, less than half of the ducks that exist lay eggs, as it is only the healthy female ducks of egg-laying age that lay eggs (37). Furthermore, there are even cases in which the predicate is true of only a tiny proportion of instances—only about 1 percent of mosquitoes carry West Nile virus (38). Why, then, is it OK to characterize these kinds in terms of predicates that apply to less than half of the instances or even a tiny minority of instances? Puzzling in a different way are cases like (39), in which the vast majority of instances have the property in question, but that is not sufficient for licensing a generic.

These observations show that it is neither necessary nor sufficient for a high proportion of instances of a kind to possess the characterizing property to license a true generic statement. If, at bottom, generics do not express quantitative generalizations, then what is the nature of the generality expressed by generic statements?

6.1.2 Hidden (and Nonhidden) Rules, Regulations, and Laws

The alternative to thinking of generics as expressing quantitative generalizations is to think of them as expressing rules, regulations, or laws that govern the behaviors of people and kinds of things more generally (Carlson, 1995). For example, "Bishops move diagonally" expresses a rule that governs how people can, and usually do, move a certain type of piece when they play chess. The existence of the rule and the truth of the generic are not based on a certain number or proportion of prior episodes of people moving bishops diagonally while playing chess, after which the rule is established or abstracted. Instead, the rule is there from the beginning and is meant to guide people's behavior in any and all episodes of chess playing that involve moving a bishop. Although consciously known rules and regulations of this sort could account for only a small proportion of the generics that we know, the thought is that other generics may be understood as expressing unconsciously known rules or laws that govern the regularities that we encounter.

Carlson (1995) argues that such a view has much to recommend it over the quantitative generalization view; however, he identifies several serious problems with the approach. These include the fact that not all true generics can be understood as directly expressing a rule or law. A weak generalization such as "Crows are smaller than ravens," which is a true generic, is unlikely to have a corresponding law or rule that makes it true. Instead, its truth likely depends on there being lawful structures that govern the size of crows and the size of ravens, which allow us to infer the true generalization about the relation that holds between them. This means that a semantics of generics will require a way of knowing both the basic set of generics whose truth depends on a corresponding rule, regulation, or law and methods for inferring indefinitely many other true generics whose truth does not directly depend on the existence of a corresponding law, but whose truth can be inferred from the generics whose truth is based on a corresponding law.

Although that is an immensely reasonable picture of what a semantics of generics ought to look like, Carlson (1995) notes some additional, and formidable, barriers to developing such a theory. To begin with, we are largely ignorant of the wide variety of mechanical, electromagnetic, chemical, genetic, biological, psychological, economic, sociological, and other laws that govern the regularities that we encounter. Furthermore, many phenomena are the product of the interaction of multiple laws. Given our general ignorance of the laws that are operative, it is unclear how we could distinguish the generics that are made true by lawful structures from those that are inferred from those more basic generics.

Other problems arise when we *do* have some idea about what underlies the truth of a generic. This is because we often have the wrong idea. So, for example, some people may believe that oak trees grow from acorns because of the genetic structure of acorns, others may think that it is because there are little oak trees inside acorns that come out when the acorn shell dissolves, yet others may think that God actively shapes each acorn into a tree, some may hold one belief at one time and a different belief at a different time, and many others may have no idea how oak trees grow from acorns.

Nevertheless, Carlson (1995, 235) says: "*Oak trees grow from acorns* is true in the mouths of all, however far-fetched their account of why it is true, or even if there is no account at all. People can still judge a generic true or false in the presence of misconceptions—or even the absence of any conceptions—of the grounding real-world conditions."

Given all of this, it seems like "our semantics must be an 'agnostic' system which need not depend on the particular propositional identity of a law" (Carlson, 1995, 235). The problem with this conclusion, he notes, is that entailment depends on propositional identity and therefore it undermines a semantics in which the truth of weak generalizations such as "Crows are smaller than ravens" is entailed by more basic generics about the size of crows and the size of ravens that may be licensed by corresponding laws.

A rules-and-regulations approach to generics that is rooted in the formal structure of commonsense kind concepts can meet many of these challenges and also shed light on various puzzling phenomena concerning generics.

6.2 Generics from the Perspective Provided by the Theory of Conceptual Form

6.2.1 Generic Knowledge That Is Intrinsic to the Conceptual Form of Kind Concepts

The theory of conceptual form seeks to explicate how the perspectives provided by lexically expressible concepts are encoded in their formal structure, which provides instructions for how to think and reason about the relations among the elements of the conceptual form of a concept. Our focus has been on lexically expressible kind concepts and the ways in which different classes of kind concepts allow us to think about kinds, instances of kinds, and the properties that characterize and distinguish a kind and its instances from other kinds of things. We have argued that, at their core, kind concepts are generative mechanisms that can generate an unlimited number of instance-of-kind representations that are numerically distinct but need not differ in their qualitative properties. The generative mechanism contains a component that represents the character that the kind intrinsically possesses and distinguishes it from other kinds. That character consists of the properties that the kind has by virtue of being the kind of thing that it is. Each such property is understood to be an aspect of being that kind of thing. Furthermore, these are properties that we think instances of the kind are supposed to have, that instances have because they are the kinds of things that they are, and that are expected to be present in instances generally. Although our focus in the previous chapters was on identifying the conceptual form of the mechanisms that encode the perspectives that we take when thinking and talking about things as instances of kinds and about kinds, the conceptual form of kind concepts encodes an important class of generic knowledge—generic knowledge that represents a principled connection between a kind and a property. That is because representing a kind requires representing a principled connection between the kind and at least some of the properties that are part of its character. This is done by specifying the values of some of the aspect representations within the conceptual form of the kind concept. Prasada and Dillingham (2009) suggest that this is done by coindexing the aspect representations (e.g., the color aspect of CANARY) with the concept that represents the value (e.g., YELLOW), as illustrated in figure 6.1. Representing generic knowledge that involves principled connections is thus an intrinsic part of the conceptual form of kind concepts.

6.2.2 Representing Nonaccidental Generalizations via Conceptual Form

Generics express nonaccidental generalizations that are not merely true of some sample of items but are expected to be true of other, as-yet-unobserved items of the relevant sort (Goodman, 1955; Dahl, 1975; Shipley, 1993; Carlson, 1995). The conceptual form of kind concepts is ideally suited for representing this way of thinking about generics. To begin with, it provides the means for generating an unlimited number of instance-of-kind representations, each of which is expected to have the character that the kind has by virtue of its being the kind of thing that it is. That character distinguishes the kind from other kinds and is understood to consist of the properties that are true of the kind by virtue of its being the kind of thing that it is. As such, the instances are understood to have those properties *because* they are the kinds of things that they are. It is thus no accident that instances of that kind have those properties. Having each of those properties is understood to be an aspect of being that kind of thing. This is the case, not just for some sample of instances of that kind, but for the unlimited number of instances that the mechanism allows us to think about. As such, the conceptual form of kind concepts provides the means for representing

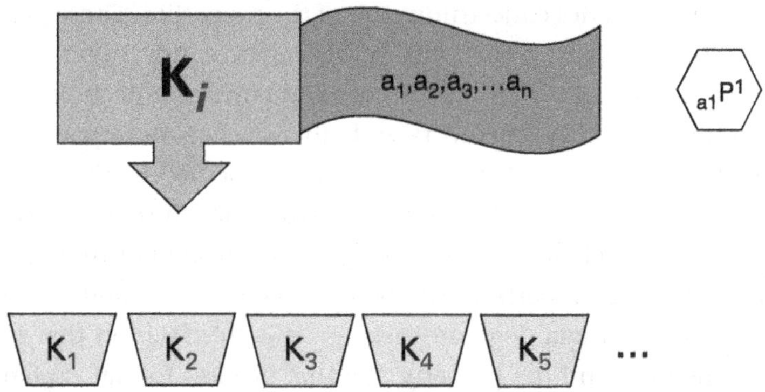

Figure 6.1
Establishing a kind concept requires specifying at least one aspect of the kind's character. It is hypothesized that this can be done by coindexing an aspect representation with the concept that specifies the value for that aspect. This diagram illustrates property P1 as specifying the value for aspect a1 by being coindexed to it. *Source*: Author.

nonaccidental generalizations concerning the character of the kind. This is because kinds cannot be distinguished merely numerically; they must also be distinguished by their character. As such, to represent a kind is to necessarily represent at least some of the properties that the kind has by virtue of its being the kind of thing that it is, which in turn are expected to be present in the unlimited number of instances of that kind, and thus to represent generic knowledge about the kind. Because the conceptual form represents both what is responsible for the generalization (the fact that the instances are the kinds of things that they are) and the fact that it is not limited to a specific sample but is true of an unlimited number of instances, it represents a nonaccidental generalization.

6.2.3 Nature of the Generic Knowledge That Is Intrinsic to the Conceptual Form of Kind Concepts

The conceptual form of kind concepts provides a perspective for thinking about kinds and their instances in which they are characterized by an intrinsic qualitative unity that consists of the properties that the kind has by virtue of its being the kind of thing that it is. These properties are all understood to be aspects of being the kind of thing in question. That unity is nonaccidental and present in an unlimited number of instances, and thus it licenses generics concerning each of the properties. Those properties are lawfully connected to the kind; however, the connection represented by conceptual form differs from the laws and lawful connections that one finds in physics, biology, chemistry and other sciences in important ways.

First, the generic knowledge intrinsically embodied in the conceptual form of kind concepts provides a perspective in which there is a nonaccidental whole-part relation between being a specific kind of thing and having certain specific properties (e.g., between being a dog and having four legs or between being a dog and barking). It is this relation that lawfully connects the kind and the property and that licenses formal explanations of the properties so connected. We can explain, for example, the four-leggedness of Fido by citing the fact that Fido is a dog because having four legs is represented as one aspect/part of being a dog (Prasada & Dillingham, 2009; Prasada et al., 2013). This form of explanation works because the existence of the whole implies the existence of the parts of the whole. This way of understanding the lawful connection between kinds and properties is purely formal. It exploits the whole-part relation between being a given kind of thing and those properties that are understood to be an aspect of being that kind of thing. It does not make reference to any laws or characteristics that derive from the material properties of the instances of the kind. Although formal explanations have often been said to be circular or tautological (Molière, 1673/2012; Hutchison, 1991; Gelman, Cimpian, & Roberts, 2018: Spaulding, 2022), and thus are proposed to be mere placeholders for explanations rather than explanations (Gelman et al., 2018), recent experiments show that formal explanations are distinguished from tautological statements (Rivera et al., 2023; Aslanov & Guerra, 2023) and are not understood to be mere placeholders for explanations but rather are understood to be explanatory (Rivera et al., 2023).

As discussed in chapter 5, classes of kinds can be formally distinguished by the number and types of properties that have a principled connection to the kind. As such, the same formal whole-part relation enters into representing kinds and licensing generics across domains. What differ from domain to domain are the number and types of properties that are aspects of that unity.

Second, the qualitative unity that is understood to characterize the kind and its instances is not only understood to be nonaccidental, but also to be *perfect,* in the sense of being complete and thus having all the parts that it is supposed to have and none that it is not supposed to have. Instances that lack one of the properties that they are supposed to have are imperfect and judged to be incomplete or have something wrong with them (Prasada & Dillingham, 2009; Prasada et al., 2013), or to be mutants (Leus, 2021). The presence of this normative element clearly distinguishes the lawful connections intrinsic to the conceptual form of kind concepts from those hypothesized by scientists. Furthermore, it makes generics that involve principled connections immune to revision on the basis of "counterexamples." The fact that there are dogs that do not have four legs cannot render the thought DOGS ARE SUPPOSED TO HAVE FOUR LEGS false. The normative does not describe how things are. The unity that characterizes the kind does not simply describe what is the case in an unlimited number of instances of the kind, but what is *supposed to be*; and thus it provides a basis for evaluating the perfection of instances of that kind. In this, it clearly departs from the types of laws and lawful connections that scientists hypothesize.

Finally, the lawful connection between the kind and a property that is understood to be part of the kind's intrinsic character supports the expectation that instances of the kind will generally possess the property. Given that the existence of a whole implies the existence of its parts, the conditions necessary for being an instance of a given kind of thing will generally be sufficient for the presence of the properties that are represented as aspects of being that kind of thing. Although it is possible that an instance will lack a property that is understood to be an aspect of being a given kind of thing if the conditions that prevent the presence of the property exist, we have reason to think that instances of the kind will generally have the property in question unless there are reasons to think that such conditions generally exist.

We thus see that the conceptual form of kind concepts provides a perspective from which kinds are understood to be lawfully connected to properties that are part of the kind's intrinsic character via a formal whole-part relation and that this way of thinking about the lawful connections differs in important ways from the lawful connections formulated by scientists to capture regularities in the world around us.

6.2.4 The Conceptual Form of Other Types of Generic Knowledge

Generic knowledge that involves principled connections is intrinsic to the conceptual form of kind representations; however, not all generic knowledge involves principled connections. Other generic knowledge involves representing causal connections or statistical connections between kinds and properties.

Generics such as "Mosquitoes carry West Nile virus" and "Pit bulls maul children" represent a causal connection between the kind and the characterizing property. The characterizing properties in such cases are not understood to be aspects of being a given kind of thing but rather to be causally connected to being that kind of thing, and thus consequences of being that kind of thing (Prasada et al., 2013). Generics that involve a causal connection, but not a principled connection to the kind, involve strikingly good or bad properties and can characterize the kind even when the property is present in a very small percentage of instances (Leslie, 2007, 2008). How often the property is realized in instances of the kind depends not only on the causal connection between the kind and the property, but also on how often enabling conditions are present and preventive conditions absent. Furthermore, properties that have a causal connection to a kind do not license normative expectations or formal explanations for the characterizing property (Prasada et al., 2013).

Generics such as "Cars have radios" and "Dogs wear collars" represent a statistical connection between the kind and the characterizing property. In such cases, the characterizing properties are understood to be neither an aspect or a consequence of being a given kind of thing, but they are nevertheless lawfully connected to the kind. The lawful connection in such cases depends on the existence and regular presence of causal mechanisms and conditions that are extrinsic to our understanding of the kind, which cause instances of the kind to have the property in question. Generics that characterize a kind in terms of a property that has a statistical connection to kind are possible only when "enough" (usually the majority of) instances of the kind have the property in question.

How, then, are generics that involve causal and statistical connections between kinds and properties represented? Given that the mechanism for representing generic knowledge involving principled connections is the mechanism that is needed to represent a kind, that mechanism must also enter into representing generic knowledge involving properties that have a causal or statistical connections to that kind. Although this may seem obvious, it highlights an asymmetry between different types of generics that has not been previously discussed. Generic knowledge involving causal and statistical connections presupposes the existence of generic knowledge involving principled connections. There could not be a kind, say "blicks," for which we know generics that involve a causal connection or a statistical connection, but no generics involving principled connections. The reason for this is that there are no mechanisms for representing bare kinds. Each kind representation comes from a class of kind representations for which there are principled connections to specific types of properties that characterize and distinguish kinds of that sort (see chapter 5). In principle, one could establish a causal connection between a kind and a property at the same time as when the kind representation is established, and thus at the same time that some generic facts involving principled connections are established, but not before that. In most cases, however, generic facts involving causal connections would be acquired only after one has acquired at least some generic knowledge involving principled connections. This should not be surprising because there are relatively few strikingly good or bad properties, and for many kinds, we likely do not know any. As such, there are relatively few generics that are represented via a causal connection to the kind only. Furthermore, although we can represent causal connections between kinds and properties for material kinds, we cannot do so for immaterial kinds (e.g., triangles), whereas we can represent principled connections for all kinds.

Because our causal knowledge is generally quite skeletal (Gelman, 1990; Keil, 2003), the representation of the causal connection must be understood in very schematic terms. For example, in representing a causal connection between MOSQUITOES and CARRY WEST NILE VIRUS, we commit to nothing more than the idea that there is *something* about mosquitoes that makes them carriers of West Nile virus when relevant enabling conditions exist and preventive conditions do not. This means that we need a placeholder representation within the kind representation that is understood to be causally connected to properties that are strikingly good or bad. Generic knowledge that is represented via a causal connection between a kind and a property is represented in figure 6.2 via an arrow between the placeholder within the kind representation and the property.

This is similar to the skeletal causal understanding that psychological essentialism attributes to natural kind concepts and some social kind concepts (Gelman, 2003; Neufeld, 2022). In fact, this placeholder can be understood as also having a causal connection to those properties that have a

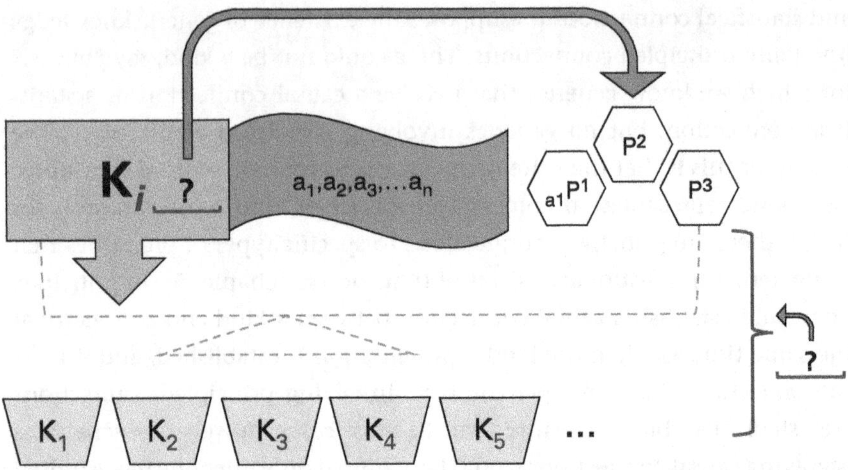

Figure 6.2
Mechanisms for representing principled, causal and statistical connections between kinds and properties. *Source*: Author.

principled connection to the kind. Prasada (2017) notes that because we generally do not know what the causal essence of any kind of thing actually is and merely have a placeholder for the essence (Gelman, 2003), the essence cannot be used to identify which properties of a kind are understood to be caused by the essence. Instead, Prasada (2017) proposes that the properties that have a principled connection to the kind are also understood to be caused by the usually unknown essence (see figure 6.3). Muradoglu, Marchak, Gelman, and Cimpian (2022) provide some evidence for that proposal by showing that formal explanations of properties (a cue that the property has a principled connection to a kind) can help children identify properties as being caused by the essence.

Generic knowledge that involves a statistical connection between the kind and the characterizing property involves properties that are understood to neither be aspects of being that kind of thing nor a consequence of being that kind of thing, but are nevertheless lawfully connected to the kind. In these cases, we assume that there is some causal mechanism or set of causal mechanisms that are extrinsic to being that kind of thing that produce the property in question in instances of the kind. The presence of this casual connection is not sufficient for licensing a generic, though. It is only

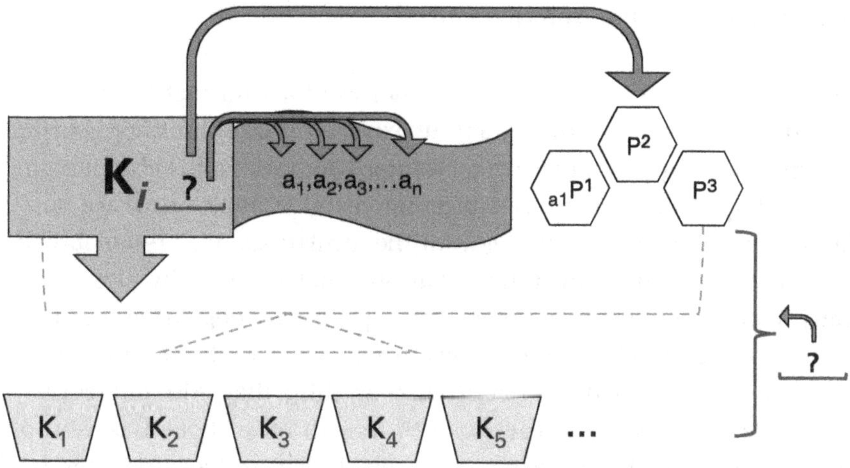

Figure 6.3
Mechanisms for representing principled, causal, and statistical connections among kinds and properties, augmented to show that the unknown essence is understood to cause the properties that the kind has principled connections to, in addition to properties that have only a causal connection to the kind (P^2). *Source*: Author.

if the causal mechanism is regularly present and able to cause "enough" instances of a kind to have a given property. When enough instances of a kind have the property in question, a statistical connection is represented between the kind and the property and a generic involving that kind and the property is licensed. Figures 6.2 and 6.3 represent a statistical connection between a kind and a property via a dashed line. Such a connection presupposes the presence of an unknown cause, represented via a placeholder and a question mark. The presupposition that enough instances have the property in question is depicted in figures 6.2 and 6.3 via the dotted triangle that extends over multiple instances. Although it is commonly assumed that a property being present in the majority of instances is sufficient for licensing a generic involving a statistical connection, it remains for future research to determine the quantitative conditions under which generics that involve statistical conditions are licensed (for some possibilities, see Cohen, 1999; Tessler & Goodman, 2019; Ryu, Yang, & Park, 2022). One interesting possibility is that what counts as enough instances may be specified by the tolerance principle (Yang, 2016).

6.3 Proportions—Puzzling No More

The variation in the proportion of instances of a kind that have a characterizing property is puzzling and unexpected if one thinks of generics as expressing a quantitative generalization. The variation is also puzzling if generics are understood as expressing lawful connections if we don't know the real-world laws that govern the regularities. We presumably do not know the lawful biological mechanisms that explain why mosquitoes carry West Nile virus or the ones that explain why dogs have four legs, or the psychological and sociological mechanisms that enter into barns generally being painted red; nevertheless we think that "Mosquitoes carry West Nile virus" is a true generic even though only about 1 percent of mosquitoes do so, but we expect virtually all dogs to have four legs and the majority of barns to be red, given the truth of "Dogs are four-legged" and "Barns are red."

The theory of conceptual form explains how we can have these statistical expectations despite being ignorant of the lawful mechanisms that govern these regularities. The conceptual form of kind concepts contains three formally distinct ways of characterizing kinds. Each represents a lawful connection between the kind and certain properties; however, as we will see next, the nature of the lawful connection and how that is related to the proportion of instances that have the property in question differs in systematic ways.

The conceptual form of generics such as "Dogs have four legs" represents a principled connection between the kind and a property, and the property is understood to be an aspect of being that kind of thing. This way of characterizing kinds is licensed by the nonaccidental whole-part relation between the kind of thing that something is and the characterizing property. Given that the conditions necessary for the existence of a whole are generally sufficient for the existence of the part/aspects of that whole, the property is expected to generally be present in instances of the kind. Only if conditions that prevent the existence of that aspect of the whole, but not the whole itself, are present will an instance of that kind fail to develop that property. Instances may also lack the property if external conditions exist that remove the property from an instance once it already has it (e.g., when a dog loses a leg in an accident). In the absence of reasons to think that these special conditions exist, though, the fact that an instance of the kind exists is a good reason to think that the instance will have the property, and thus we expect instances to generally have those properties that are understood to be aspects of being that kind of thing.

Generics such as "Mosquitoes carry West Nile virus," on the other hand, are licensed when the conceptual form of the kind concept represents a causal connection between the kind of thing and the characterizing property. In such generics, the characterizing property is understood not to be an aspect of being that kind of thing. Because the property is not understood to be an aspect of being that kind of thing, the existence of an instance does not provide a reason to think that the conditions necessary for the presence of the property in an instance hold. The causal connection licenses the generic; however, whether the property is present in instances of the kind also depends on the presence of enabling conditions and the absence of preventive conditions. Consequently, generics such as "Mosquitoes carry West Nile virus" do not license any expectations concerning how often instances will have the characterizing property. The prevalence is determined by the frequency with which the enabling conditions are present and preventative conditions absent.

Finally, the conceptual form of kind representations can license generics such as "Dogs wear collars" by representing a nonaccidental statistical connection between the kind and a property when enough (usually a majority) of instances have the property *and* there is assumed to be a mechanism or set of mechanisms that causes that property. It is the presumed existence of this extrinsic cause that allows us to treat the presence of the property in instances of the kind to be nonaccidental, and thus to expect that unobserved instances will also have the property in question. What proportion of instances must be known to have a property before a statistical connection between a kind and property can be represented will have to be determined by future research. For this class of generics, unlike generics involving the other two types of connection, that proportion is a necessary condition for the licensing of the generic.

We thus see that the theory of conceptual form helps explain why generics differ in how prevalent their characterizing properties are. The distinct types of lawful connections between kinds and properties that can be represented by the conceptual form of kind concepts support different expectations concerning the prevalence of characterizing properties in instances of the kind. Two facts about prevalence that aren't explained by the conceptual form of kind concepts illustrated in figure 6.2 are the licensing of minority characteristic generics such as "Ducks lay eggs" and "Lions have manes," as well as why high prevalence is insufficient for licensing generics like "#Human beings are right-handed." I turn now to describing a recent elaboration of the conceptual form of principled connections that addresses both of those issues.

6.3.1 Representation of Minority Characteristic Generics

Minority characteristic generics such as "Lions have manes" appear to be represented via principled connections between the kind and the characterizing property. Prasada et al. (2013) found that participants judge the property to be an aspect of being that kind of thing, endorse formal explanations of the property (e.g., "That animal has a mane because it is a lion"), and endorse normative expectations concerning the property (e.g., "Lions are supposed to have manes"). These findings suggest that generics like "Lions have manes" are represented via the same type of connection as generics like "Lions have tails." One problem with this conclusion is that whereas we expect lions generally to have tails, we do not think that lions generally have manes—we expect only male lions to have manes. Another problem is that whereas we think that there is something wrong with a lion that does not have a tail, we do not think that there is anything wrong with a lion that doesn't have a mane if it is a female lion. The normative expectation holds only for male lions.

In fact, some have suggested that although minority characteristic generics such as "Lions have manes" appear to characterize a kind, they are implicitly understood as characterizing only a relevant subset of the kind (e.g., Declerck, 1991; Pelletier & Asher, 1997; Asher & Pelletier, 2012; Lazaridou-Chatzigoga, Stockall, & Katsos, 2019). So, for example, "Lions have manes" is implicitly understood as "Male lions have manes." There are at least three problems with that proposal. First, if the statement is really about an appropriate subtype, why is the statement about the kind licensed at all? Why don't we simply say, "Male lions have manes"? Second, it is unclear how the proposal could be limited in the proper way. If what appear to be statements characterizing a kind are readily understood as characterizing a subset of the kind, what prevents us from accepting "Birds are pink" as a way of saying "Flamingos are pink"? Finally, Prasada and Leus (2024) point out that the proposal implies that we represent a lawful connection between only a subset of the kind (e.g., male lions) and the property in question (e.g., manes), and thus that we do not also represent a lawful connection between the rest of the kind (e.g., female lions) and the property (e.g., manes); however, that seems wrong. We are not agnostic about the relation between female lions and manes. Instead, we think there is a principled connection between being a female lion and the *absence* of a mane. Female lions are supposed to *not* have manes. And the absence of a mane in a female lion can be explained by pointing out the fact that it is a female lion.

Minority characteristic generics are challenging because the perspective provided by kind concepts is one in which properties that have a principled connection to the kind are understood to be properties that are supposed to be shared by all instances of the kind, and the differences among instances of the kind are understood to be accidental and randomly distributed. This assumption is violated by the properties that characterize kinds in minority characteristic generics as nonarbitrary subsets of instances of the kind displaying nonaccidental, principled differences in their possession of those properties. For kinds that reproduce sexually, there are principled differences between male and female instances with respect to the parts and processes needed for reproduction, and sometimes the properties that identify them as male or female instances (e.g., manes). Furthermore, it appears that the presence of a property in one sex implies its absence in the other sex. That is, the principled differences are of a very specific sort. First, they are necessary for the continued existence of the kind. Second, they involve a difference in the mode of existence of a property that characterizes both sexes rather than a nonmodal property difference between the two sexes. Finally, although many minority characteristic generics assume a principled division between instances of two types, with the two types being complementary and exhaustive of instances of a kind (e.g., male/female lions), this need not be the case. For example, we think that the generic "Bees make honey" is true even though it is only female worker bees that do so—male bees (drones) and queen bees do not produce honey. Thus, the property characterizes instances of one type, and the remaining instances are characterized by the property's absence whether those other instances constitute a type or not.

Based on these observations and the finding by Prasada et al. (2013) that minority characteristic generics involve principled generics, Prasada and Leus (2024) proposed the *principled within-kind variation hypothesis*,

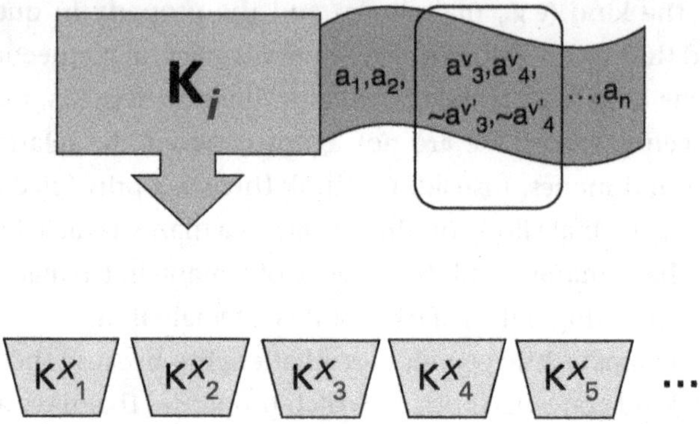

Figure 6.4
Conceptual form of kind concepts that provides the means of representing principled within-kind variation. The top part provides the basis for representing majority characteristic generics such as "Lions roar" and minority characteristic generics such as "Lions have manes." The bottom part has the instance-of-kind representations that it generates. *Source*: Author.

according to which minority characteristic generics are licensed when principled within-kind variation is necessary for the kind to exist. In such cases, a principled connection between the characterizing property and the kind is represented as a principled connection between the presence of the characterizing property and the instance type that it characterizes, as well as a principled connection between the property's absence and the rest of the instances of the kind.

Prasada and Leus (2024) note that the hypothesis requires that our understanding of the conceptual form of kind concepts and the manner in which principled connections are represented need to be modified in three ways. First, the component that represents the qualitative character of the kind must include some aspects of being that kind of thing that show principled within-kind variation. These properties must be marked with a formal feature that identifies the instance type for which there is a principled connection for the presence of that property. Properties that show principled within-kind variation are identified by the superscript V in figure 6.4.

Second, for each property for which there is principled variation, there must also be a representation of the absence of that property that is formally marked to identify it as characterizing the complement of the instance type identified by the property. In figure 6.4, this is done via the superscript V′. Instances that are not of the instance type V are thus understood to have a principled connection for the absence of that property. This means that properties that show principled within-kind variation (e.g., having manes for lions), like those that do not (e.g., roaring for lions), have a principled connection to the kind and thus enter into characterizing all instances of the kind. Properties that show principled within-kind variation have a principled connection for the *presence* of the property for instances of the relevant instance type and a principled connection for its *absence* for other instances of the kind. On the other hand, properties that do not show principled within-kind variation have a principled connection for the presence of the property for all instances of the kind.

The third change required for representing principled within-kind variation is a mechanism for identifying the instances that possess a property that shows principled within-kind variation as being the relevant instance type. In figure 6.3, the instance type of instances is identified via the superscript variable *x* on the instance-of-kind representations whose value is determined by any property that displays principled within-kind variation that is present in that instance.

Some evidence in favor of the principled within-kind variability hypothesis, and thus the conceptual form in figure 6.4, is presented by Prasada and Leus (2024). In one experiment, they showed that formal explanations of the presence of the characterizing property were rated as sounding natural for formal explanations that referenced the kind (e.g., "That has a mane because it is a lion") and the instance type linked to the property (e.g., "That has a mane because it is a male lion"), but not in terms of the instance type not linked to the property (e.g., "#That has a mane because it is a female lion"). A second experiment showed that normative expectations for the characterizing property followed a similar pattern. Normative expectations for the presence of the characterizing property were rated as sounding natural when the expectation was about the kind (e.g., "Lions are supposed to have a mane") or the instance type linked to the property (e.g., "Male lions are supposed to have manes"), but not when the expectation was about the instance type not linked to the property (e.g., "#Female lions are supposed to have manes"). Finally, a third experiment showed that participants have universal normative expectations for instances of the item type characterized by the property (e.g., "All male lions are supposed to have manes"), but allow that not all instances may have the property and thus are less likely to endorse the corresponding descriptive statement (e.g., "All male lions have manes"). On the other hand, they rejected the corresponding universal normative expectation and the universal descriptive statement when it involved the kind or instances of the item type not linked to the property. As such, participants had expectations concerning the prevalence of the property that derives from the normative dimension of principled connections and those statistical expectations display principled within-kind variation.

Together, the experiments reported by Prasada and Leus (2024) show that minority characteristic generics represent a principled connection between the characterizing property and the kind such that there is a principled connection between the *presence* of the characterizing property and the instance type that it characterizes, and there is a principled connection between the property's *absence* and the rest of the instances of the kind.

Given the revised form of principled connections illustrated in figure 6.3, we see how the conceptual form of kind concepts can account for the proportion of instances that have the characterizing properties of minority characteristic generics.

6.3.2 Why Human Beings Cannot Be Characterized as Being Right-Handed

The revisions to conceptual form required for accommodating minority characteristic generics can also help explain why, in certain cases, highly prevalent properties do not license generics that characterize the kind in terms of that property. Why, for example, can't we characterize human beings as right-handed even though the vast majority of human beings are right-handed? Previous research on generics has distinguished properties like EXTINCT or INVENTED, which apply to kinds but not instances of kinds, from properties like FURRY and FRIENDLY, which can apply to instances of kinds or characterize kinds (Krifka et al., 1995). Accommodating minority characteristic generics points to a third class of properties such as MALE, FEMALE, QUEEN, and RIGHT-HANDED, which apply to instances but cannot be applied to kinds. Such properties apply to only a principled subset of the instances within a kind, and therefore they imply the existence of instances within the kind that *necessarily and nonaccidentally* lack that property. As such, they could not possibly apply to all instances of the kind. In this, they differ from properties like FURRY, FRIENDLY, or CARRY WEST NILE VIRUS, all of which could, in principle, be true of all instances of the kinds that they characterize, even if, in fact, they are not true of some, and perhaps the vast majority, of instances of the kind. Because properties such as male, female, and right-handed are understood as applying only to a principled subset of instances of a kind, they cannot characterize the kind (Prasada & Leus, 2024). We thus see that the conceptual form of kind concepts can explain why despite the high prevalence of right-handedness in human beings, human beings cannot be characterized as being right-handed.

The conceptual form of kind concepts helps (1) explain why generics vary in the proportion of instances that have a characterizing property, (2) identify the factors that determine the proportions for generics involving different connection types, and (3) explain why certain predicates cannot characterize kinds even if they are highly prevalent. The variation in the proportion of instances that have the characterizing property in true generics is puzzling no more.

6.4 The Limited "Agnosticism" of Conceptual Form

As discussed earlier in this chapter, Carlson (1995, 235) suggests that "our semantics must be an 'agnostic' system which need not depend on the particular propositional identity of a law" because generics are understood to be true by people that have very different ideas about what law makes the generic true, as well as by people who have no idea about what law makes the generic true. The theory of conceptual form shows how our representation of generic knowledge can be agnostic in the manner intended, but it also highlights some important limitations of the "agnosticism" metaphor.

Let us begin with the conceptual form of generics that involve a causal connection, such as "Mosquitoes carry West Nile virus." Recall that such generics involve representing a causal connection between a placeholder representation within the kind and a property that is understood to be strikingly good or bad. The form of the representation captures the fact that we understand that there is *something* about the kind of thing that causes the property when enabling conditions are present and preventive conditions absent. As the representation does not specify what it is about the kind that causes the characterizing property, on its own, it allows for the possibility that the generic is understood to be true while also having almost no idea of the lawful connection that makes it true.

The qualifier "almost" is important. Representing the generic in this manner requires that there is *something about that kind of thing* that is lawfully connected to the property in question. This is different from the conceptual form underlying generics that involve a statistical connection, such as "Barns are red." For generics of that sort, the conceptual form contains a placeholder that is *extrinsic to the kind* that represents the causal mechanisms that are responsible for regularly producing the property in enough instances of the kind. Thus we see an important limitation on the agnosticism that we have concerning what makes generics like "Mosquitoes carry West Nile virus" and "Barns are red" true. We know that in the first case, it is something that is intrinsic to the kind, and in the second case, it is something that is extrinsic to the kind that causes instances of the kind to have the property in question (when relevant conditions hold). As such, it is not the case that we have *no* conception of what makes a generic true, and thus that we are not completely agnostic about what makes a generic true.

Because the conceptual form of generics involving causal connections and those involving statistical connections contain placeholders and represent causal connections in schematic terms without further specifying the nature of the causal mechanisms involved, they allow for different specifications of the contents of the placeholders, as well as different specifications of the causal mechanisms. As such, they allow the generics to be "true in the mouths of all, however, far-fetched their account of why it is true" (Carlson, 1995, 235). This is true because the conceptual form of generics that involve causal connections and those that involve statistical connections is *underspecified.*

Turning our attention to the conceptual form of generics such as "Dogs have four legs," which involve principled connections, a different picture emerges. For these generics, we understand instances of the kind to have the characterizing property because they are the kind of thing that they are and having that property is represented as one aspect of being that kind of thing. So, for example, we understand Fido to have four legs because he is a dog, and having four legs is one aspect of being a dog. Notice that our understanding of why "Dogs have four legs" is true does not make reference to a placeholder. Furthermore, formal explanations for properties that have a principled connection to a kind are generally endorsed by participants (Prasada & Dillingham, 2006, 2009; Prasada et al, 2013; Rivera et al., 2023), as is the idea that these properties are understood to be aspects of being that kind of thing (Prasada & Dillingham, 2009; Prasada et al., 2013).

The conceptual form of DOG, which is largely shared by different individuals, ensures that we are generally in agreement, and not agnostic, about what makes generics such as "Dogs have four legs" true by representing having four legs as an aspect of being a dog. On the other hand, the conceptual form of DOG only ensures that we are in agreement that *something* about being a dog causes a dog to be a potential carrier of rabies. This allows us to be agnostic about what that something is or disagree about what that something is. Similarly, the conceptual form of DOG only ensures that we are in agreement that *something* extrinsic to being a dog regularly causes dogs to wear collars, and thus we can be agnostic about what that something is or disagree about what that something is.

Because conceptual form embodies both a formal mode of understanding and a causal mode of understanding, it allows us to represent both a formal aspect-kind relation between characterizing properties that have a principled connection to the kind and a causal connection between the kind and the property for material kind concepts such as DOG (but not immaterial kind concepts such as TRIANGLE). This means that although we know that "Dogs have four legs" is true because of the lawful part-whole relation between the characterizing property and the kind, we also know that there is a lawful causal connection between being a dog and having four legs, and thus that there is something about being a dog that causes it to have four legs. This allows us to think and reason about this regularity from a perspective from which we know exactly what makes it true, as well as a perspective from which we know what makes the generic true in only a very underspecified way and thus can be largely agnostic of what makes it true.

We thus see how the conceptual form of kind concepts captures the ways in which we are both agnostic and not agnostic about the lawful connections that make a generic true.

6.5 Generating Weak Generalizations

The limited agnosticism of conceptual form also helps address a problem that a truly agnostic semantic system generates. Carlson (1995) notes that if our semantics is an "agnostic" system, which does not depend on the particular propositional identity of the laws which make generics true, one cannot generate weak generalizations via entailment because the entailments would depend on the propositional identity of the laws that make the generics from which weak generalizations are inferred true.

The conceptual form of kind concepts avoids this problematic state of affairs because although it allows people to "judge a generic as true or false in the presence of misconceptions," it does *not* allow them to do so in the "absence of any conceptions—of the grounding real world conditions" (Carlson, 1995, 255). As discussed previously, generics that involve a causal connection between the kind and the characterizing property represent a highly underspecified conception of the grounding real-world conditions that make the generic true. It is something about that kind of thing that is causally connected to the property in question. Similarly, generics that involve a statistical connection between the kind and the characterizing property also represent a highly underspecified conception of the grounding real-world conditions that make the generic true. It is something that is extrinsic to being that kind of thing that is regularly present and causes instances to have the property. Importantly, it is not the case that we have *no* conception of what the grounding real-world conditions are. In both cases, we have some conception of what they are, but that conception is severely unspecified. That is why different people can specify it in different ways and specify it in ways that are incorrect.

Furthermore, in the case of generics that involve principled connections, we do know the formal grounding real-world conditions, and we have an underspecified representation of the causal grounding real-world conditions. The conceptual form of kind concepts thus implicitly represents the propositional identity of the laws that render the generics represented in the conceptual form of a kind concept true. As such, the conceptual form of kind concepts implicitly represents propositions that identify a lawful formal connection (40), an underspecified causal connection (41), or an underspecified causal connection that underlies a statistical connection (42):

(40) Dogs have four legs because having four legs is an aspect of being a dog.

(41) Something about dogs causes them to have four legs.

(42) Something that is extrinsic to dogs regularly causes them to wear collars.

Weak generalizations can then be inferred on the basis of generics encoded in conceptual form. So, for example, we can infer "Spiders have more legs than dogs" from "Spiders have eight legs" and "Dogs have four legs";

"Cherries are the same color as barns" from "Cherries are red" and "Barns are red"; "Ravens weigh more than crows" from "Ravens weigh about 2 pounds" and "Crows weigh about 1 pound"; and so on . . .

Although much more would need to be said about the details of the inferential machinery and how it uses conceptual form to generate weak generalizations, I hope that what we have discussed so far makes it clear that the theory of conceptual form can capture the ways in which we can be agnostic about the real-world laws that ground the truth of generics, but also that there are limits on that agnosticism. The conceptual form of kind concepts identifies three ways in which real-world conditions can ground the truth of generics. In doing so, it also provides the basis for inferring indefinitely many weak generalizations on the basis of the generics that are encoded in the conceptual form of kind concepts, but that need not have real-world laws that directly ground them. As such, the theory of conceptual form provides a solution to one of the significant challenges faced by approaches that view generics as expressing rules, regulations, and lawful connections.

6.6 Multiple Ways of Talking (and Not Talking) About Kinds

Most languages provide multiple ways of talking about kinds of things. English has three common ways of doing so—via the use of bare plural noun phrases as in (34)–(39), via an indefinite singular noun phrase as in (43), and via a definite singular noun phrase as in (44):

(43) A dog has four legs.

(44) The dog has four legs.

(45) Dogs have four legs.

So we can characterize dogs as having four legs in three different ways using (43), (44), or (35), repeated here as (45). It is important to note that there is no unique morphosyntactic way of signaling a generic interpretation of a noun phrase, so bare plurals and indefinite singular and definite singular noun phrases can also be used in nongeneric utterances (46)–(48). Generic interpretations depend on the form of the noun phrase, as well as information about tense, aspect, predicate, and discourse factors.

(46) Dogs were barking in my backyard last night.

(47) A dog was barking.

(48) The dog was barking.

In line with the fact that generic interpretations rely on multiple sources of information is the fact that there are different constraints on the generic use of bare plural, indefinite singular, and definite singular noun phrases. The bare plural can be used to talk about any kind and any property that characterizes the kind (49). In contrast, whereas the indefinite singular can be used to characterize any kind, it cannot be used to characterize the kind in terms of any property (50). Finally, the definite singular is restricted with respect to both the properties that support generic interpretation (51) and the kinds that can be characterized through its use (52).

(49a) Dogs have four legs.

(49b) Dogs wear collars.

(49c) Dogs carry rabies.

(49d) Brown dogs have loud barks.

(50a) A dog has four legs.

(50b) #A dog wears a collar. (on a generic interpretation)

(50c) #A dog carries rabies. (on a generic interpretation)

(51a) The dog has four legs.

(51b) #The dog wears a collar. (on a generic interpretation)

(51c) #The dog carries rabies. (on a generic interpretation)

(52a) The collie has a loud bark.

(52b) #The brown dog has a loud bark. (on a generic interpretation)

Questions thus arise as to which properties can appear in generic uses of indefinite singular and definite singular generics. Are they the same or different? And which kinds can appear in definite singular generics? Further questions arise about how those constraints on kinds and properties may or may not be related to the nongeneric interpretations that the noun phrases support. Is it an accident that the indefinite singular and definite singular noun phrases have the constraints that they do on their use in generics? Could there be a language in which the nongeneric interpretation of indefinite and definite noun phrases is the same as in English, but the constraints on kinds and properties that appear in their generic uses are swapped? And finally, why are there different constraints on the expression of a general fact at all?

6.6.1 The Conceptual Form of Kind Concepts and Constraints on the Generic Use of Noun Phrases

The conceptual form of kind concepts provides the means for characterizing kinds in three ways—via properties that have a principled connection, a causal connection, or a statistical connection to the kind.

Bare plural generics can be used to characterize kinds in terms of each of the three connection types (49a)–(49c) (Prasada et al., 2013). Furthermore, bare plural generics are not limited to being used with so-called well-established kinds that we think and talk about using lexically expressible kind concepts such as DOG; they can also be used with what we have been calling ad hoc subkind concepts, such as BROWN DOG (49d), which are the result of conceptual combination of a kind concept (e.g., DOG) with another concept (e.g., BROWN) that creates an ad hoc subkind concept (e.g., BROWN DOG). As discussed in chapter 5, ad hoc subkind concepts have a formal structure that does not project an aspect structure that could represent the intrinsic character of the ad hoc subkind. As such, properties that characterize the ad hoc subkind cannot be understood as having a principled connection to the subkind. Because the form of ad hoc subkind concepts allows concepts that have no intrinsic connection to one another to be combined, the instances of the ad hoc subkind concept cannot be assumed to have a principled qualitative unity, as is the case for lexically expressible kind concepts that do not require conceptual combination. Ad hoc subkinds may nevertheless be characterized via properties that have a statistical connection (49d) or causal connection to the subkind (Prasada, 2010, 2012). Any statistical or causal connection between the ad hoc subkind and a property must be based on observation of instances of the ad hoc subkind or inferred on the basis of our world knowledge about the properties and kinds involved because the formal structure of ad hoc subkind concepts allows, but does not encode, any intrinsic statistical or causal characteristics of the ad hoc kind. Because instances of an ad hoc subkind (e.g., BROWN DOG) are necessarily instances of the kind that the ad hoc subkind is derived from (e.g., DOG), we have a reason to think that the ad hoc kind can be characterized in the same way as the kind it is derived from, in the absence of reasons to the contrary. For example, we know that dogs bark, wear collars, and carry rabies, and so, unless we have reasons to think otherwise, we have reason to think that brown dogs bark, wear collars, and carry rabies and do so at prevalence levels that are the same as for dogs generally. Note that although properties that have a principled connection to a kind can be assumed to characterize ad hoc subkinds derived from that kind, they are understood as having a statistical rather than a principled connection to the ad hoc kind. Thus, although we may expect brown dogs to bark, we think that they bark because they are dogs, not because they are brown dogs. Similarly, we think that they are supposed to bark because they are dogs, not because they are brown dogs. We thus see how the conceptual form of kind concepts enters into the interpretation of bare plural generics.

Indefinite singular generics sound natural if a kind is being characterized by a property that has a principled connection to the kind (50a), but not if the kind is being characterized by a property that has a statistical connection to the kind (50b) (Prasada & Dillingham, 2009) or a property that has a causal connection to the kind (50c) (Leslie, Khemlani, Prasada, & Glucksberg, 2009). Those findings fit well with the suggestion in Lawler (1973) that indefinite singular generics require the property to be essential and the observation by Cohen (2001) that indefinite singular generics carry normative force. Principled connections specify the sense in which such properties may be thought to be essential (Carlson, 2010; Prasada & Dillingham, 2006) and they also carry normative force. Thus, it appears that indefinite singular generics can be used to characterize a kind if the characterizing property has a principled connection to the kind. This constraint on the interpretation of indefinite singular generics is, of course, readily stateable in terms of the conceptual form of kind concepts.

Indefinite singular generics can also be used to characterize ad hoc subkinds that are represented via phrasal concepts such as GREEN BOTTLE (53):

(53) A green bottle (usually) has a narrow neck (Krifka et al., 1995).

(54) A collie has a loud bark.

In such cases, the characterizing property *cannot* have a principled connection to the ad hoc subkind because ad hoc subkind concepts are generated via conceptual combination and the phrasal concept does not project an aspect structure (see chapter 5). Such generics must be understood as expressing what we think is likely to be true of an arbitrarily chosen instance of the ad hoc subkind represented via the phrasal concept. As such, they seem to require an explicit quantificational element such as "usually"/"sometimes"/"rarely" and sound quite unnatural without such an element. This contrasts with cases in which indefinite singular generics are used to characterize a kind (50a) or a non–ad hoc subkind (54) where no explicit quantificational element is required and it is understood that all instances have the characterizing property by virtue of being the kind of thing that they are, although they may lack it for reasons other than their being that kind of thing. Conceptual form thus enters into determining which properties can characterize an ad hoc subkind, as well as specifying how indefinite singular generics are interpreted differently when characterizing kinds and ad hoc subkinds.[1]

Although there is much less empirical work on definite singular generics, such generics sound natural when they are used to characterize kinds in terms of properties that have a principled connection to the kind (51a), but not when the property has a statistical (51b) or causal connection (51c) to the kind (Leslie et al., 2009). As with indefinite singular generics, this constraint on the use of definite singular noun phrases to characterize kinds is naturally stated in terms of the conceptual form of kind representations.

Definite singular noun phrases are also limited in the kinds that they can characterize (Dayal, 2004; Krifka et al., 1995). Definite singular noun phrases can be used to characterize "well-established kinds" like DOG (51a) and COLLIE (52a), but not non–"well-established kinds" such as BROWN DOG (52b). This restriction is unsurprising, given the requirement that the characterizing properties of definite singular generics need to have a principled connection to the kind. As discussed previously, the conceptual form of ad hoc subkind representations such as BROWN DOG does not allow for the possibility of representing principled connections between ad hoc subkinds and properties. Of course, the conceptual form of ad hoc subkinds does not rule out the possibility of characterizing them in terms of properties that have a statistical connection to the ad hoc subkind (49d). Why, then, can we not use a definite singular noun phrase to characterize an ad hoc subkind in terms of a property that is statistically prevalent among its instances (52)?

(55) #The brown dog has a loud bark.

Definite singular generics seem not to be possible for ad hoc subkinds, no matter the connection type of the characterizing property. Why? A potential clue lies in another constraint on the kinds that can be characterized via definite singular generics. Carlson (2010) reports the observation by Vendler (1971) that some kind terms seem to be "too general" to appear in definite singular generics (56c). The impossibility of (56c) shows that being a "well-established kind" cannot be sufficient for licensing a definite singular generic:

(56a) Animals can move on their own.

(56b) An animal can move on its own.

(56c) #The animal can move on its own. (on a generic interpretation)

Is this a further constraint on the use of definite singular generics, or might the constraint against "too general" kind terms and ad hoc subkinds derive from a single constraint? The theory of conceptual form suggests the latter. Specifically, it appears that definite singular noun phrases can be used generically when they characterize what we understand to be a single kind of thing. Concepts such as DOG and COLLIE provide such a perspective. Although both also allow a perspective in which they contain instances of different kinds (of dogs/collies), we need not adopt such a perspective. We can think of their instances as simply being one kind of thing. This is not possible with ad hoc kind concepts such as BROWN DOG. Such concepts provide the means for thinking about dogs of *any* kind that are brown.

Evidence for this difference in the perspectives provided by subkind concepts and ad hoc subkind concepts comes from an experiment reported by Prasada, Hennefield, and Otap (2012). They found that subkind concepts such as COLLIE are understood to be a single kind of thing to a greater extent than ad hoc subkind concepts such as BROWN DOG, and also that ad hoc subkind concepts such as BROWN DOG are understood as not necessarily a single kind of thing to a greater extent than subkind concepts such as COLLIE are. As discussed in chapter 5, the conceptual form of superordinate kind concepts such as ANIMAL contain instances only indirectly—as instances of the subkinds that are directly contained in the kind. As such, instances of such superordinate kinds cannot be assumed to be of a single kind. Thus, the constraint against non–"well-established kinds" such as BROWN DOG and the constraint against "too general" kind concepts such as ANIMAL appearing in definite singular generics is a single constraint against any type of kind concept that has a conceptual form preventing it from assuming that the instances that it is used to think and talk about constitute a single kind of thing. Or, to put it positively, generic interpretations of definite singular generics noun phrases require the noun phrase to be mapped to a kind concept that has a conceptual form that allows the instances of the kind to be assumed to be a single kind of thing.

In sum, definite singular generics require the characterizing property to have a principled connection to the kind, and the noun phrase must identify a kind concept whose instances must be thought to be a single kind of thing. The conceptual form of kind concepts is crucial for capturing both constraints.

6.6.2 Identifying Nonaccidental Connections Between Kinds and Properties and Constraints on Generic Expressions

As we have seen, conceptual form enters into capturing the constraints that govern the ways in which indefinite singular, definite singular, and bare plural noun phrases can be used to express generic thoughts. But why are there different constraints on the use of different linguistic devices for expressing generic thoughts? Why isn't there just one linguistic device? Or multiple linguistic devices, all with the same constraints? Relatedly, is there a relation between the nongeneric interpretation of the noun phrases and the constraints that we find on the generic interpretation of the noun phrase? I'd like to suggest that the variation that we see is not arbitrary but instead reflects three rational forms of generalization that are implicit in commonsense conception. Two of those forms of generalization are also explicitly used by scientists and mathematicians when they seek to identify nonaccidental connections among things.

As scientists, we are constantly in search of lawful connections among various aspects of reality. To do so, we need to distinguish those things that are lawfully connected from those that happen to be connected by mere chance. Statistics provide an essential tool for distinguishing accidental generalizations that appear by chance from nonaccidental generalizations that are the result of lawful connections. The intuition behind the form of generalization licensed by statistics is the idea that, as Aristotle put it, things that happen always (or for the most part) do not happen by chance. So things that are connected in a sample of a population should not be found to be so connected if we look at more and more samples from the same population if their connection is accidental. Similarly, accidental connections should be less likely to be found the bigger the sample we observe. Statistics provides us with formal tools for making this mode of generalization from the properties of a sample to properties of the population precise. It uses purely quantitative measures of the objects of study. As such, this mode of generalization allows for the possibility of identifying lawful connections for any and all kind of things and properties whose lawful connections we may seek to discover. It crucially depends on observations from multiple instances, and generally the more observations, the better. Furthermore, though, multiple observations are required, the identity of the instances observed is not important, and, ideally, the observations are of randomly selected samples. Domain knowledge is important for identifying what to count or measure; however, the statistical tools that are crucial to distinguishing lawful and accidental connections are content neutral. Generics expressed via bare plural noun phrases are similarly constraint free. The bare plural can be used to express nonaccidental generalizations for any kind of thing and properties of any type. It is also interesting to note that this way of talking about kinds involves an expression that is plural and indefinite. This expression can be used to talk about an actual multiplicity of instances whose identity and numerosity is left unspecified (e.g., "Dogs were barking in my backyard last night"), or to characterize a kind in terms of what is true of a sample of instances of unspecified instances of unspecified numerosity. Bare plural noun phrases thus make available a representation that in statistics can play the role of a randomly selected sample of arbitrary size. As such, they make available a representation that is relevant to a content-neutral mode of generalization.

Another mode of generalization is common in mathematics. Mathematicians do not arrive at the conclusion that triangles have internal angles that add to 180 degrees by measuring the angles of a sample of triangles and performing statistics on the likelihood of the sample's properties as having occurred by chance. Instead, they seek to determine what can be concluded about an arbitrary instance of a category by applying abstract formal principles that are not specific to the category in order to derive conclusions that are specific to the category. An arbitrary instance is not an arbitrarily chosen instance of the kind, but instead a representation that is a cognitive tool that allows us to reason about the category as a whole. An arbitrary instance represents characteristics of the category that are shared by all instances of the category, and none of the properties that are specific to individual instances of the category, and thus are not shared by all instances of the category. For example, the triangle in figure 6.5 leaves unspecified the values of the lengths of each of the line segments that make up the sides of the triangle, as well as the angles at which the line segments meet.

To derive a conclusion about the internal angles of a triangle, mathematicians extend the line segments to be lines so a triangle can be understood

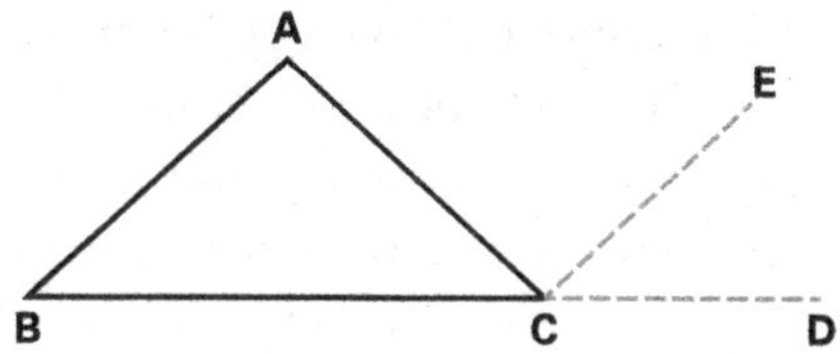

Figure 6.5
An image used to prove that the sum of the internal angles of all triangles is 180°.
Source: Author.

to be the closed figure that results from three mutually intersecting lines. Doing this allows us to bring our knowledge about intersecting lines to bear on triangles and their internal angles. Next, a line that is not part of the triangle, but is parallel to one of the sides, is introduced. Our knowledge of intersecting lines allows us to conclude that ∠ECD = ∠ABC and ∠ACE = ∠CAB. Given that the three angles at C add up to 180° and are equivalent to the internal angles of the triangle, we can conclude that the internal angles of the triangle add up to 180°. Because the reasoning was carried out on the representation of an arbitrary triangle that does not contain any information about the lengths of the sides or the size of the angles, the conclusion was arrived at independent of that specifying information and applies to any and all triangles.

Unlike the statistical mode of generalization, generalization via the consideration of an arbitrary instance can happen by considering a single instance as an arbitrary instance; however, this mode of generalization requires that we have abstract formal knowledge of a more general category that can be applied to reason about the properties of the specific category of interest. As discussed in chapter 5, the theory of conceptual form generates classes of kind concepts that are distinguished by their formal characteristics, including classes that are distinguished by the types of properties to which kinds in the class have principled connections. The knowledge embodied in the formal structure of these classes of kind representations can be used to acquire knowledge about specific kinds within the class on the basis of experience with only a single instance. Knowledge with a single instance is sufficient because the value that the single instance displays on a dimension that is determined by the kind is understood to have a lawful connection to the kind and the other dimensions determined by the kind, independent of the values displayed by the instance of the specific kind being considered. This mode of generalization is limited to properties that have principled connection to the kind, as we can have abstract knowledge of how properties of that sort are related to one another and the kind independent of how they are realized by any specific kind within the class. Generalizations that involve a causal connection to a kind cannot be acquired in this manner because which properties have a causal connection to a kind depend on the specific kind of material that a given kind of thing is constituted of and thus cannot be known independent of what a specific kind of thing is constituted of. Properties that have only a statistical connection to a kind depend on the properties of the instances of specific kinds, and thus the statistical connection between the kind and the property type cannot be known independent of the specific kind.

Generics expressed via an indefinite singular noun phrase are limited in just this manner. They can be used to express generalizations that involve a principled connection to the kind, but not a statistical or causal connection to the kind. It is also interesting to note that this way of talking about kinds involves an expression that is singular and indefinite. Indefinite singular noun phrases can be used to talk about a single instance of a kind whose identity is not specified in a definite manner (e.g., "A dog was barking in my backyard last night"), but it can also be used to characterize the kind by specifying what is a true or an arbitrary instance. An arbitrary instance is a tool of cognition, which represents a single instance of a kind without specifying the properties that are specific to any instance. Thus, in both its generic and nongeneric use, the noun phrase makes the same information available and leaves the same information unspecified. In one case, it is interpreted as making reference to an instance of the kind; in the other case, it is understood as providing the means for characterizing the kind in terms of properties that have a principled connection to the kind.

Finally, there is a mode of generalization that is even more constrained, which is a specialization of the mode of generalization that uses arbitrary instances. Like that mode, it is limited to properties for which there are principled connections to the kind because it is possible to know which types of properties have a principled connection to kinds within a given domain independent of knowledge of specific kinds within the domain. However, unlike that mode of generalization, which is available, in principle, for kinds at multiple levels of abstraction, the present mode of generalization is limited to kinds for which we can assume that there is no principled variation in the properties determined by the kind. Given these limitations, we can engage a "seen one, seen them all" mode of thinking. We can think of a specific instance as representative of the kind because we know that there is no systematic variation in the properties of instances of that kind. The variation that there is, is mere noise. This mode of generalization, which we may call the *representative* mode of generalization, works for kind representations at the basic level and more specific levels. Thus, there is an important sense in which one can sensibly say that if you know the specific properties that a particular dog, raccoon, or horse has on dimensions that have a principled connection to the kind for animate beings, which you know what all dogs, raccoons, or horses are like on those dimensions, but the same is not true for ANIMAL. If I know the specific properties that a particular kind of animal has because it is an animal, I cannot reasonably assume that all animals or all kinds of animals have those specific characteristics. Other kinds of animals will realize the properties that animals have by virtue of being animals in other ways. In sum, if we limit ourselves to considering kinds that have no principled variation and we consider only properties on dimensions for which things of that kind have principled connections, then we can observe a single instance of the kind and take the specific values that it has on those dimensions and assume that all other instances of the kind should have those values.

Generics expressed via a definite singular noun phrase are limited in the manner that the representative mode of generalization is. They can be used to express generalizations that involve a principled connection to the kind, but they can do so only when considering kinds for which no principled variation is expected, and thus all instances can be seen as instances of one kind with only random variation due to noise. Note that this way of talking about kinds involves an expression that is singular and definite. Definite singular noun phrases can be used to talk about a single instance of a kind whose identity is known (e.g., "The dog was barking in my backyard last night" must involve a known dog), but it can also be used to characterize the kind by thinking of that instance as representative of the kind.[2]

It thus appears that we have at least three rationally motivated modes of generalization, each of which has different constraints on its range of applicability. It also appears to be the case that these three modes of generalization are reflected in three different ways in which English allows us to characterize kinds of things. The constraints that appear on the range of the modes of generalization are also reflected in the ways in which bare plural, indefinite singular, and definite singular noun phrases can be used when characterizing kinds. Finally, it appears that the form of the noun phrases is systematically related to the mode of generalization that they support. In each case, the definiteness and number of the noun phrase when being used in nongeneric utterances are also relevant to thinking and talking about the mechanism in terms of which a given mode of generalization characterizes kinds (i.e., in terms of a sample, an arbitrary instance, or a representative instance). It thus appears that much of the variability found in the ways in which we can characterize kinds is not arbitrary but derives from three modes of generalization that are implicit in commonsense conception. This is consistent with the idea that what is systematic about generic phenomena is to be found at the conceptual level (Collins, 2015). It will be important to investigate the extent to which this systematicity shows up cross-linguistically. Examining other linguistic devices in English and the constraints on their generic uses will also be important.

6.7 Generics and Conceptual Form

Generic knowledge and generic language pose difficult challenges for both psychologists and linguists. The theory of conceptual form provides a new perspective from which we can approach many of those challenges. In this chapter, I have sought to illustrate how the perspective provided by conceptual form can (1) explain why the proportion of instances that have a characterizing property in true generics varies, and why they vary in the ways that they do; (2) provide a satisfactory account of how minority characteristic generics such as "Lions have manes" are represented; (3) provide an account for why high prevalence of a property is not sufficient for licensing a generic in certain cases like "#Human beings are right-handed"; (4) provide a theory of the cognitive mechanisms needed to represent the rules, regulations, and lawful connections thought to license generics; (5) explain how generics can be represented in a manner such that we are largely (but not completely) agnostic about what makes a generic true and thus accommodate the fact that people can differ about what makes a generic true, or may have (almost) no conception of what makes a generic true; (6) explain how weak generalizations can be inferred from the basic generics that can be represented in the conceptual form of kind concepts; (7) capture the known constraints on the use of different ways of characterizing kinds in English; (8) provide a unified constraint on the generic use of definite singular generics that explains why they cannot be used with non–"well-established" kinds such as BROWN DOG or with "well-established" kinds like ANIMAL; (9) explain why there are multiple ways of linguistically characterizing kinds; (10) explain why different ways of expressing generic facts have the constraints they do; and (11) explain how the form of noun phrases is related to both the mode of generalization underlying a particular generic and the nongeneric interpretation of that noun phrase.

Although many questions remain unanswered and challenges unmet, I hope to have made the case that approaching those questions and challenges from the perspective provided by the theory of conceptual form is a worthwhile enterprise.

7 Explaining Systematic Polysemy

Chapter 6 showed how the theory of conceptual form sheds light on a wide range of phenomena concerning generic knowledge and generic language. In this chapter, I show how the theory of conceptual form can explain why names for kinds of things, across languages, display certain patterns of systematic polysemy but not other patterns (Srinivasan & Rabagliati, 2015). This fact has been noted, but there has been no satisfactory account of it to date.

7.1 Forms of Polysemy

If we consider the meanings of the word "bat," for example, it is evident that it has two clearly distinguished meanings that are completely unrelated to one another. It is an accident that the piece of sports equipment used to hit a ball has the same name as a nocturnal animal. Similarly, there is no connection between the meaning of the word "bank" that allows us to talk about a financial institution and the meaning that allows us to talk about an area next to a body of water. Such words are called *homonyms*. Homonyms have two or more unrelated meanings, but a single phonological form. In other cases, the connection between multiple meanings of a word is not completely accidental. For example, "line" can be used to talk about a geometric entity, as well as the thing that we stand in when we are waiting for something, and also a bit of dialogue in a play (among other things). In this, and other such cases, the multiple senses of the word appear to be related by similarity or to involve analogical extensions of meaning. Such words are said to be *polysemous*. Polysemous words have two or more meanings that are related but a single phonological form. In most cases of polysemy, there is no systematic pattern of meaning extension. So, for example, it is not the case that names for geometric entities (e.g., "angle") can generally also be used to talk about formations of people waiting to do something, or bits of dialogue in a play. Thus, whereas "line" is polysemous and the multiple meanings are not accidentally related to one another, the pattern of polysemy is not systematic. There are other cases, however, where the multiple meanings of a word have an intrinsic and systematic connection to one another. In these cases, there are *classes* of words that show the same patterns of polysemy. For example, not only can we use "bottle" to talk about a container as well as the contents of the container (57), we can do the same for "box," "carton," "pot," and many other names for containers and their contents:

(57) The bottle broke. / The bottle spilled on the floor.

(58) A potato fell on the floor. / Would you like some potato?

(59) The magazine had a stain on it. / The magazine was shocking.

(60) Hurley knocked on the door. / Hurley walked through the door.

(61) The chicken was clucking. / The chicken was undercooked.

(62) The school is very inclusive. / The school burned down.

Similarly, although (58) illustrates two related meanings of "potato," the same related pair of meanings can be found for "eggplant," "zucchini," "banana," and a whole range of names for fruits and vegetables. The alternation in the meanings of "magazine" in (59) are also found for "newspaper," "book," "encyclopedia," and the names for many other types of reading materials. The pair of meanings in (60) is also found for "window." "Chicken" and other names of animal kinds (e.g., "turkey," "fish," or "lamb") can be used to talk about the meat of those animals (61). And "school" and many other names of institutions (e.g., "bank," "church," or "prison") can also be used to talk about the buildings in which they are housed (62). All these cases display *systematic polysemy* (Apresjan, 1974; Pustejovsky, 1995), as they involve patterns of polysemy that are found for whole classes of words. Patterns of systematic polysemy tend to be found cross-linguistically (Srinivasan & Rabagliati, 2015).

Polysemy has been the subject of much research over the past three decades as researchers have sought to discover whether there are separate entries for the different senses of polysemous words or if they are accessed via a single entry (e.g., Frisson, 2015; Foraker & Murphy, 2012; Klein & Murphy, 2001;

Klepousniotou, 2002; MacGregor et al., 2015), as well as whether, when, and how distinct senses of a word can be combined (e.g., Collins, 2017; Lohr & Michel, 2022; Murphy, 2019; 2021; Ortega-Andres & Vicente, 2019; Pietroski, 2018), whether different senses of a word are acquired together (e.g., Srinivasan & Snedeker, 2011, 2014; Zhu, 2021), and the formal character of the lexical representations of polysemous words (e.g., Pustejovsky, 1995; Asher, 2011; Quilty-Dunn, 2021). All that work takes the patterns of polysemy as given and investigates the manner in which lexical items and their senses are represented, stored, accessed, and acquired and how they can and cannot be combined.

What that body of research does *not* do is to try to explain why certain patterns of systematic polysemy are found but others are not. Why, for example, can we use "magazine" and "newspaper" to talk about the organizations that produce the associated reading material ("The magazine/newspaper hired an editor"), but we cannot use "book" or "letter" to talk about the person who wrote them ("#The book/letter hired a publicist")? Or why we cannot use the name of a vehicle to refer to its operator ("#The bus announced the next stop") or the other way around ("#The bus driver was very crowded"), even though there are systematic and lawful connections between the classes of vehicles and their operators (Ritchie & Prasada, 2024)?

A notable exception in this line of research is Rabagliati, Marcus, and Pylkannen (2011). They investigated whether the acceptability of patterns of systematic polysemy is related to conceptual centrality, cue validity, salience, or the similarity of senses. They found that these characteristics of conceptual structure were "surprisingly poor predictors" of the acceptability of systematically related senses (Rabagliati et al., 2011, 485). More recently, Ortega-Andrés and Vicente (2019) have indicated that other conceptual factors may explain patterns of systematic polysemy. They suggest that patterns of systematic polysemy may be due to coactivation packages in which the senses are connected via explanatory relations including "realization, actualization or implementation" (14). Although such notions are central to our conceptual representations, it is not clear that they can explain which patterns of systematic polysemy occur and which do not. For example, vehicles of all kinds require an operator to achieve their functions, and yet we cannot use the term for a vehicle to refer to its operator, or vice versa (Ritchie & Prasada, 2024).

7.2 The Role of Conceptual Form in Explaining Patterns of Systematic Polysemy

We can explain why certain patterns of systematic polysemy occur and why others do not via a core characteristic of the conceptual form of lexically expressible kind concepts. You may recall from chapter 5 that classes of kind concepts can be formally distinguished by what they individuate. As we will see, those classes display different patterns of systematic polysemy because the name for a kind can also be used to talk about the things that the kind individuates. Patterns of systematic polysemy concerning kind/instance-of-kind/subkind-of-kind senses have generally not been discussed in the literature (but see Dölling, 1995); however, they are intrinsic to kind representations. As such, it is plausible that an individuation relation may also be required for licensing *other* polysemous patterns of kind terms. If so, patterns of systematic polysemy may be accounted for via an *Individuation for Polysemy* principle, which may be formulated in the following way: "When a representation of type A is required for the individuation of a representation of type B, the name that we use to identify representations of type A and the things they are used to talk about can be used to identify representations of type B and the things that they are used to talk about." In what follows, I will show how the Individuation for Polysemy principle can account for many patterns of systematic polysemy, including those found in (57)–(62). It can also explain why systematic polysemy is not found when there is no individuation relation between representations of type A and representations of type B, even if they bear other salient non-accidental conceptual relations to one another.

This explanation of why some patterns of systematic polysemy occur and others do not is essentially the one that Katherine Ritchie and I proposed (Ritchie & Prasada, 2024). My presentation of it here differs in scope and detail, as I show how the account can explain patterns of systematic polysemy beyond those considered by Ritchie and Prasada (2024).[1]

7.3 Individuation and Patterns of Polysemy Within the Conceptual Form of Lexically Expressible Kind Concepts

The conceptual form of different classes of lexically expressible kind concepts outlined in chapter 5 support different patterns of systematic polysemy. Those formal structures were motivated by a range of conceptual and linguistic phenomena, including some concerning individuation and some concerning polysemy. Here, we reconsider those conceptual forms with a focus on how they involve individuation and license different patterns of polysemy. I will discuss the distinct classes of kind concepts using a single clear example from each of those classes to focus on how the conceptual forms from distinct classes of kind representations underlie different patterns of polysemy in those classes.

7.3.1 Class of Kind Concepts That Individuate Instance-of-Kind and Subkind-of Kind Representations

We begin by considering the conceptual form of kind concepts from the class that DOG belongs to (see figure 5.1, repeated here as figure 7.1a). Here briefly is a recap of the formal structure of kind concepts in that class (for further details and explication, see chapter 4). As you may recall, kinds are represented via generative mechanisms that are capable of generating an unlimited number of instance-of-kind representations. Furthermore, any two kind representations (say, DOG and CAT) are distinguished via an index that captures the numerical distinctness and oneness of each of the represented kinds. Of course, kinds cannot be merely numerically distinct.

They also differ in their characters. That way of distinguishing kinds is captured by the component of the kind representation that represents the character of the kind via the properties that have principled connections to the kind (see figure 7.1b). Kind representations are thus distinguished from one another in two ways—via an index that represents each kind as one kind that is numerically distinct from other kinds, and via their character, which is represented via the properties that have principled connections to the kind.

Each kind representation can generate an unlimited number of instance-of-kind representations. Those instance-of-kind representations do not exist and cannot be identified independent of the kind representations that generate them. The instance-of-kind representations have two components: one that identifies the kind of thing each instance is understood to be and an index that distinguishes it from all other instance-of-kind representations for that kind (see figure 7.1c). The index is generated by the kind mechanism and is devoid of any intrinsic content. As such, instance-of-kind representations are understood to be merely numerically distinct. The index distinguishes the instance-of-kind representations and

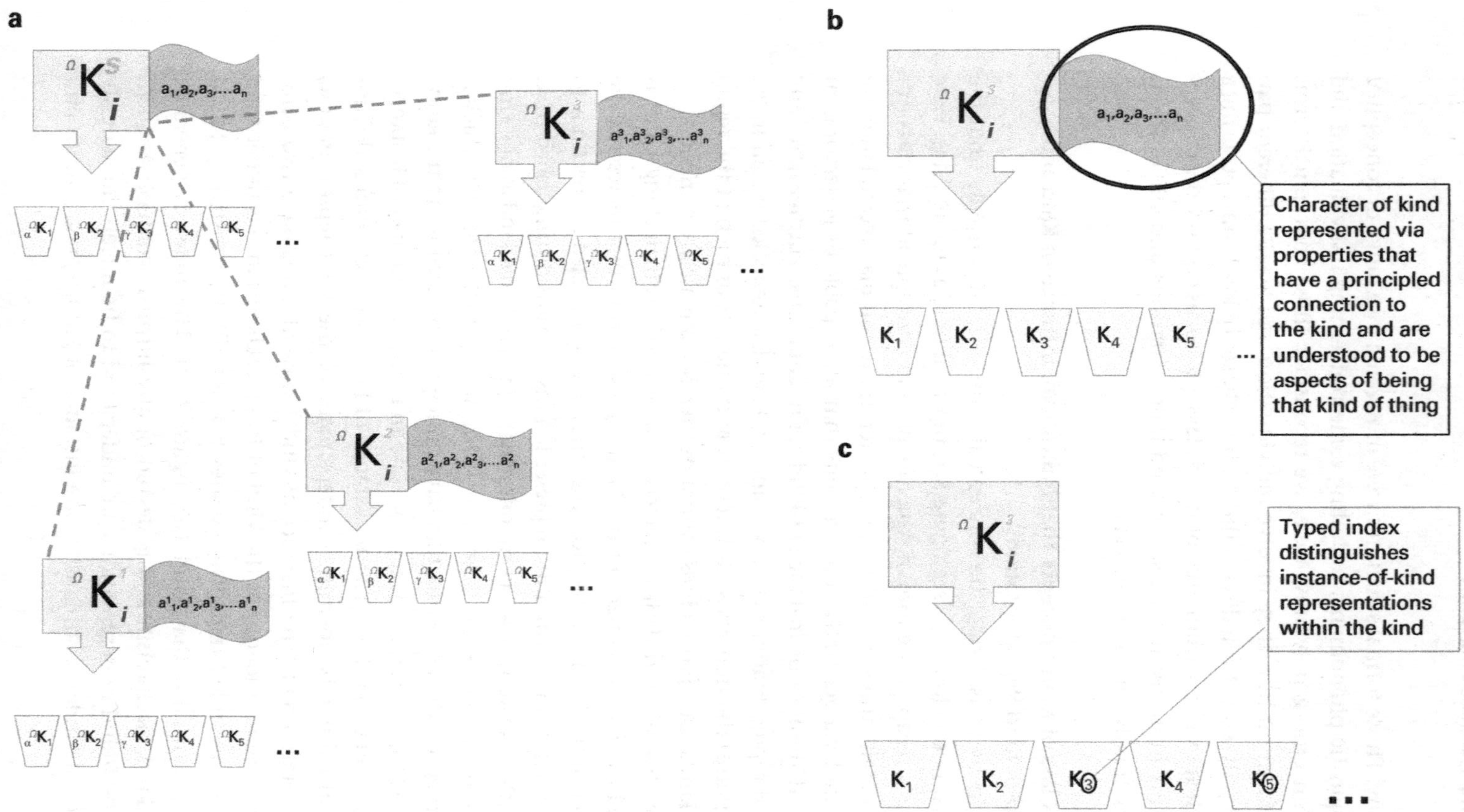

Figure 7.1

(a) The conceptual form of kind concepts that intrinsically contain both instances and subkinds. (b) The component that represents the character of the kind via principled connections to properties that are understood to be aspects of being that kind of thing. (c) The typed index that distinguishes instances within the kind and represents each instance as one instance of that kind. *Source*: Author.

represents the quantitative equivalence of the instance-of-kind representations. Each counts as one instance-of-kind representation of that kind, and thus instance-of-kind is a unit of quantification for that kind. That unit of quantification cannot be specified independent of the kind representation.

The kind representation can also be used to generate subkind representations. The subkind representations do not exist and cannot be identified independent of the kind representation that generates them. Unlike instance-of-kind representations, which can be merely numerically distinct, subkind representations cannot be merely numerically distinct but must also differ in character. The characters of the subkinds are understood to be distinct ways of realizing the character of the kind of which they are subkinds. The subkind representations are numerically distinguished from one another via an index that is generated by the kind representation. The index represents each subkind as one subkind of the kind. As such, the subkind representations are quantitatively equivalent, and thus subkind-of-the kind is a unit of quantification. That unit of quantification cannot be specified independent of the kind.

In sum, kind representations within this class individuate both instance-of-kind and subkind representations and thus provide two units of quantification whose identity and existence depends on the kind representation. As such, expressions that identify some number (*N*) of units of the kind (e.g., "I like *N* dogs") may be understood as either being about *N* instances of that kind, or as being about *N* subkinds of that kind.

We thus see that there is a relation between individuation and the use of the name of the kind to talk about instances and subkinds of that kind. This is because in identifying a kind, we also identify how instances and/or subkinds are distinguished, and thus what counts as one instance and/or one subkind of that kind. Because the kind representation determines what counts as one instance/subkind of the kind in question, the name for the kind can be used to identify the kind, as well as to identify any number of its instances and subkinds.

That connection between individuation and the form of systematic polysemy displayed by the name of the kind can be highlighted by considering the contrast between the class of lexically expressible kind concepts that we just considered and another class of lexically expressible kind concepts that differs in what the kinds in that class individuate and the pattern of polysemy that the names of the kinds display.

7.3.2 Class of Kind Concepts That Individuate Subkind-of Kind Representations, but Not Instance-of-Kind Representations

If we consider concepts from the class that contains kind concepts such as SAND, we see that the kind concept can be used to generate subkind representations, but it does not, on its own, generate instance-of-kind representations. The kind specifies how subkinds are distinguished and the sense in which they are one. But we can generate instance-of-kind representations only in combination with another concept such as PILE, PORTION, or CUP, which must be either explicitly combined with the kind representation or contextually inferred. Such additional concepts specify the means for distinguishing and identifying instances of the kind that they modify as one. As such, the kind representation provides only one unit of quantification that provides the basis for quantifying subkinds of that kind. Expressions that identify some number of units of the kind (e.g., "I like *N* sands") must be understood as being about that number of subkinds of that kind. They cannot be understood as being about some number of instances of that kind unless we infer a unit of quantification from the context that is not determined by the kind.

The quantification over subkind interpretation is available across contexts because the unit of quantification is provided by the kind representation and not dependent on context. In contrast, the quantification over instance interpretation will vary with context because what to count as an instance will vary from context to context. Furthermore, there may be situations in which it is not obvious what to count as an instance and the quantification over instance interpretation may simply not be available.

In sum, kind concepts in this class individuate subkind-of-kind representations, but not instance-of-kind representations, and the name for the kind can also be used to talk about subkinds of that kind, but not to talk about instances of the kind (without the help of additional concepts and context). As such, the pattern of polysemy found in this class is predicted by the Individuation for Polysemy principle.

7.3.3 Class of Kind Concepts That Individuate Instance-of-Kind Representations, but Not Subkind-of Kind Representations

A third class of kind concepts includes concepts like ROOK. Kind representations in this class individuate instance-of-kind representations, but not subkind-of-kind representations. As such, the kind provides only one unit of quantification that provides the basis for quantifying instances of that kind. Expressions that identify some number of units of the kind (e.g., "I like *N* rooks") must be understood as being about that number of instances of that kind rather than that number of subkinds. Quantification over subkinds requires explicit mention that the unit of quantification is the subkind (e.g., "I like two *kinds of* rooks"), or the unit of quantification must be inferred to be subkinds on the basis of contextual information and thus usually will require stage-setting. No stage-setting or explicit identification of instances as the unit of quantification is required for understanding the sentence as quantifying over instances.

The kind representation ROOK is a clear example of this class. The reason for that is that the properties that the kind has by virtue of being that kind of thing are stipulated (i.e., "Rooks are chess pieces that are stipulated to have a certain form and move any number of squares horizontally or vertically"). As such, there can't be systematically different ways of realizing those characteristics. The scope of the class of kind concepts that individuates instance-of-kind representations, but not subkind representations clearly covers all kind concepts where there is an arbitrary relation stipulated between form and function. However, it may be that the class includes all artifacts even though we clearly know subkinds for many artifact kinds (e.g., desks, nightstands, and coffee tables are all kinds of tables). Sentences like "I like two tables: dining tables and coffee tables" sound odd even though "I like two kinds of tables: dining tables and coffee tables" sounds fine. This suggests that TABLE may not provide a subkind as a unit of quantification, and thus that the relation between TABLE and COFFEE TABLE may be represented differently than that between DOG and COLLIE.

Three reasons to think that is the case are the following: First, it seems like one cannot have the representation COLLIE without identifying it as a subkind of DOG, but one can characterize a DESK or COFFEE TABLE as an artifact with certain characteristics without identifying it as a subkind of TABLE. Second, whereas collies and dachshunds are understood to be simply one of a number of ways of being a dog, coffee tables and picnic tables do not seem to be simply different ways of being a table—they have their own reasons for existence. Finally, whereas all dogs, and all subkinds of dogs, are understood to come into existence via the same process, and thus to have a principled unity, different kinds of tables are understood to come into existence via different processes, suggesting that they do not have the same type of principled unity which they realize in different ways. Systematic empirical research is needed to determine the robustness of these intuitions and the exact scope of the class of kind representations that individuate instance-of-kind representations, but not subkind representations. We are currently investigating these questions, and Fenton and Prasada (2025) report some experimental data suggesting that whereas representations of kinds of living things (e.g., DOG) individuate subkinds (e.g., COLLIE), representations of artifact kinds (e.g. TABLE) do not individuate putative subkinds (e.g., COFFEE TABLE).

7.3.4 Class of Kind Concepts That Individuate Instance-of-Kind Representations in Two Ways

Finally, a fourth class of kind concepts includes concepts like BOOK. Kind representations in this class individuate two types of instance-of-kind representations, but no subkind representations. Concrete instance-of-kind representations are numerically distinct and their qualitative differences are understood to be accidental. On the other hand, abstract instance-of-kind representations cannot be merely numerically distinct. They also have a nonaccidental abstract character that distinguishes and identifies them as one abstract instance of the kind. Each abstract instance is intelligible, in and of itself, and in a different way than every other abstract instance of that kind. Finally, each abstract instance is expected to be realized in an unspecified number of concrete instances. As such, kind representations in this class specify two units of quantification that provide the basis for quantifying in terms of concrete and abstract instances of that kind. Consequently, expressions that identify some number of units of the kind (e.g., "I like *N* books") can be understood as being about that number of concrete instances (e.g., "I like the book on the desk and the one on the shelf") or abstract instances of that kind (e.g., "I like *Syntactic Structures* and *Of Mice and Men*"). Quantification over subkinds requires explicit mention of the unit of quantification (e.g., "I like two kinds of books—recipe books and history books"), or the unit of quantification must be inferred to be subkinds of that kind on the basis of contextual information and thus will usually require stage-setting.

We thus see across four classes of kind representations that the name that we use to identify kind representations within a class can be used to identify the instance-of-kind/subkind-of-kind representations individuated by the kind representation, as well as to talk about the instances/subkinds represented by those representations. On the other hand, we cannot use the name used to identify kind representations to talk about those instance-of-kind/subkind-of-kind representations that are not individuated by the kind representation. This shows that an individuation relation is crucial for licensing patterns of polysemy that involve the use of a noun to talk about kinds, as well as concrete and abstract instances and/or subkinds that are understood to be intrinsic to that kind and whose representations are generated and individuated by the kind representation. In short, the patterns of polysemy found and not found across these four classes of kind concepts illustrate the validity of the Individuation for Polysemy principle.

7.4 Individuation and Patterns of Polysemy Across Kinds

We turn now to explicating how the Individuation for Polysemy principle can account for patterns of polysemy in which the name for one kind of thing can be used to talk about a *different* kind of thing and its instances. What we will see is that there is a class of kind concepts that includes kinds of organizations that are individuated by the kinds of things that they produce. This class includes, among others, organizations that produce magazines and organizations that produce newspapers. For members of this class, we can use the name of the product that that kind of organization produces to talk about that kind of organization and its instances. Other kinds of organizations are not individuated by what they produce, and, as we will see, we cannot use the name of the kind of things that such organizations produce to talk about those kinds of organizations.

Let us consider magazines and the organizations that exist for the sake of producing magazines. Obviously, organizations that exist for the sake of producing magazines do not have an existence independent of magazines, and magazines do not simply pop into existence but are produced by organizations. As such, the kinds are mutually dependent on one another in different ways. What about the formal properties of their kind representations? What, if any, formal relations do the kind representations MAGAZINE and OPM (the kind representation of organizations that produce magazines)[2] have to one another?

Given that OPM is a kind representation, it has an index that represents it as one kind that is numerically distinct from other kinds. Furthermore, like all kind representations, OPM has a component that represents its character, which characterizes and distinguishes it from other kinds. That component will represent the fact that the organization exists for the sake of producing magazines. Similarly, MAGAZINE has an index that represents it as one kind that is numerically distinct from other kinds and has a component that represents its character, which characterizes and distinguishes it from other kinds.

How is the index for MAGAZINE related to the index for OPM? A moment's reflection makes it clear that that which distinguishes magazines from other kinds of things also distinguishes the kind of organization that exists for the sake of producing magazines from other kinds of organizations. The indices for the two kind representations indicate the same thing. The index that identifies one kind of reading material and distinguishes it from other kinds of things also identifies one kind of organization that exists for the sake of producing that kind of reading material, and it also distinguishes it from other kinds of organizations. As such, MAGAZINE individuates OPM. Furthermore, because the word "magazine" functions as the index for MAGAZINE, we have an explanation for why "magazine" can be used to talk about both that kind of reading material and that kind of organization. Thus, we can say "Magazines are more popular than books" to talk about the kind of reading material, and "Magazines are having a hard time attracting quality writers and illustrators" to talk about the organizations that produce the reading material.

The instance-of-kind representations generated by OPM are also individuated by MAGAZINE. The indices that distinguish and identify the instance-of-kind representations of OPM (the different organizations that exist for the sake of producing different magazines) are determined by the indices that distinguish and identify abstract instance-of-kind representations of MAGAZINE (e.g., *Vogue, Spin*). Each instance-of-kind representation of OPM produces a different magazine in its abstract publication sense. For example, one instance-of-OPM is understood as existing for the sake of producing *Vogue*, another to produce *Spin*, and so on.

Because MAGAZINE enters into individuating both the kind of organization that produces that kind of reading material and its instances, the name for MAGAZINE, "magazine," can be used to talk about both the kind of organization that produces magazines (e.g., "Magazines are having a hard time attracting quality writers and illustrators") and the instances of that kind of organization (e.g., "That magazine hired a new editor"). Furthermore, the proper name (index) of the abstract instances of MAGAZINE can be used to talk about the organizations that are instances of OPMS (e.g., "*Vogue* hired a new editor") because the abstract instances of MAGAZINE individuate the organizations that are instances of OPMS.

These facts provide further evidence for the Individuation for Polysemy principle and show that it holds even when one kind representation individuates a *different* kind representation and its instances. Further evidence for the principle comes from two types of cases in which the kind representation does *not* individuate the producer of the kind, and thus does not license the use of the name for the kind to talk about the producer of the kind.

The first involves cases like BOOK, which like MAGAZINE and NEWSPAPER is a kind representation for a kind of reading material; however, unlike those kinds, the producer of a book is not understood to be an organization; rather, it is conceived to be an individual author (or more than one author). Authors are mentally represented as existing independent of the books that they write, and their oneness and distinctness from other people (or authors) is not dependent on the books that they write. As such, BOOK does not play a role in individuating the producer, and thus we should not be able to use "book" to talk about the producer/author. This prediction is borne out by the example that it makes no sense to say something like "#The book hired a publicist."

Second are cases in which an organization produces a product but the organization is not conceived of as being individuated by what they produce. In such cases, the organization is not represented as simply existing for the sake of producing that product but for some other reason (such as making money for its owners). In that case, the product is represented as a means to some other end. For such organizations, the product is not represented as being necessary for individuating the company, which could choose to produce other things and remain the same company. For example, a shoe company may start producing things other than shoes (e.g., sportswear, sports equipment) and may even stop producing shoes and remain the same company. As such, organizations of this kind (companies) cannot be individuated by what they produce. Nor can the instances of this sort of organization be individuated by the products they produce. For example, different shoe companies can produce the same shoe designs and thus cannot be individuated by the kinds of shoes that they produce. Given that neither this kind of organization nor its instances are individuated by what they produce, the Individuation for Polysemy principle predicts that it should not be possible to use the name for the kind of product that they produce to talk about either the kind of organization that produces it or the kind's instances. This is exactly what we find (e.g., "#Shoes are doing well on the stock market right now."/"#That shoe hired a new manager").

7.5 Individuation and Patterns of Polysemy Across Kinds and Conceptual Relations

We just saw that when the producer of a given kind of thing is individuated by the kind, the name for the kind can be used to talk about both the producer kind and its instances. In this section, we discuss cases in which the individuating concept is not the product. What we will see is that the Individuation for Polysemy principle holds for different conceptual relations between the individuating and individuated concepts.

First, consider the pattern of systematic polysemy that involves the use of the name for a kind of animal to talk about the edible matter (meat) that comes from that kind of animal. In such cases, the meat is causally dependent on the animal that it comes from. The meat may be thought to be the product; however, unlike the cases of systematic polysemy just considered where the product individuates the (producer) kind that it is lawfully related to, meat does not individuate the (source) kind that it is lawfully related to. The individuation relation runs in the other direction. Kind representations for kinds of edible matter that come from an animal kind (EMA) are distinguished from one another via indices that identify them and indicate that which distinguishes them from other kinds of EMAS. Those indices depend on the indices that individuate the animal kind representations. Kinds of meat are individuated by the kinds of animals that they come from. As such, the names for kinds of animals can be used to talk about the kinds of meat that come from them.

A parallel account can be given for the use of names of fruits and vegetables to talk about the foodstuffs that are derived from them. In both the meat and foodstuffs cases, the kind representations are representations of kinds of edible substances, and thus they do not individuate their instances—what counts as one instance of a specific kind of meat or a specific kind of fruit or vegetable stuff is not determined by the kind representation of that kind of stuff. As such, the animal and plant kinds from which they derive cannot individuate their instances, which explains why we can't say things like "#John took two eggplants" to mean that John took two portions of eggplant foodstuff.

A different form of systematic polysemy involves the use of the name of an original artist to talk about the kind that consists of that artist's works (e.g., "Picassos/Kahlos rarely sell for less than a million dollars"), as well as the individual works of art by that artist (e.g., "Two Picassos/Kahlos were damaged yesterday"). We cannot do this for the artworks produced by nonartists like myself (e.g., "#Prasadas are unlikely to sell for more than a dollar"), or even professional artists who are skilled but whose work lacks an original and distinctive aesthetic sensibility (Isaac, 2015).

The Individuation for Polysemy principle provides an account of these facts. We do not form kind representations such as PRASADA or SMITH that contain artworks by nonartists and unoriginal artists because kind representations must differ both numerically and in their character; however, the artworks of unoriginal artists differ from one another in accidental ways, so there is no particular character that distinguishes Smith's paintings from those of Prasada's. In contrast, the works of original artists such as Pablo Picasso and Frida Kahlo are understood to be not just numerically distinct, but to be characterized by the distinctive aesthetic sensibilities of their creators, and thus their artworks can be represented via distinct kind representations (e.g., PICASSO, KAHLO). The character that is understood to individuate these kinds and to be displayed by their instances is understood to be determined by the character that distinguishes the representations of the individual artists qua artists.

Picasso and Kahlo, for example, are not simply numerically distinct artists—we understand them to have distinctive and original aesthetic sensibilities that are displayed in their artworks. As such, that which individuates Picasso and Kahlo as original artists is what individuates the kinds that are constituted by their artworks and allows us to use the name of the original artist to talk about their artworks. This is also why we cannot use the name of an original artist to talk about nonartistic works that they may have produced (e.g., model airplanes made from a kit; "#That Picasso took Picasso fifteen minutes to assemble from a kit"). Those products do not display the distinctive aesthetic sensibility of the creator and differ from products of that sort (e.g., model airplanes) that are made by other producers in accidental ways, and thus such products are not individuated by their producers.

We thus see that the Individuation for Polysemy principle can also explain cases in which the individuation relation is between an individual of a specific kind (e.g., original artist) and a kind that is a product (e.g., the artist's artwork), not just cases in which the individuation relation is between two kinds or between a single kind and its instances and/or subkinds.

7.6 Contextually Determined Individuation and Patterns of Polysemy

The systematic polysemy involved in container-content alternations such as (57) involve an individuation relation between an instance of a specific kind of container and an ad hoc instance of the content type at a specific time. Thus, "bottle" can be used to talk about an ad hoc instance of JUICE in a circumstance in which a bottle individuates the portion of juice contained in it at that time, but "bottle" cannot be generally used to talk about juice. Furthermore, on other occasions, "bottle" may be used to talk about a portion of milk, or some other liquid, that is delimited as one instance by a bottle. More generally, containers, just by virtue of being containers, are capable of individuating ad hoc instances of unindividuated stuff, but what portion of which kind of stuff is individuated by an instance of what type of container varies from circumstance to circumstance. When an instance of a given container type is individuating an instance of a given type of stuff, the name for the container type can be used to talk about the individuated instance (portion) of stuff, but only at that time. As such, this form of systematic polysemy relies on the context-independent capacity that instances of container kinds possess for individuating instances of unindividuated stuffs of any kind, with the polysemous use being licensed only in those contexts in which the individuation relation is realized. We thus see that the Individuation for Polysemy principle can apply to cases in which the individuation relation is between instances of two kinds of things, which holds in a context-dependent manner.

An even more contextually determined individuation relation is relevant to licensing sentences like "The three ham sandwiches want their checks" in which "ham sandwich" is used to talk about three people who have ordered a ham sandwich and can be used to talk about ham sandwich orderers generally (e.g., "Ham sandwiches usually tip well"), or even to distinguish that ad hoc kind from other ad hoc kinds and their instances (e.g.,

"Ham sandwiches usually tip better than grilled cheeses, but this grilled cheese is a very good tipper"). One can imagine such talk between servers at a restaurant. Crucially, in this context, ad hoc kinds and their instances are being individuated on the basis of contingent relations between foods and persons. Such uses are not licensed when the word being used to talk about a person does not individuate that person. For example, "The ham sandwich wants his check" is not felicitous in a context in which all the customers ordered a ham sandwich, as HAM SANDWICH cannot provide individuating information in that circumstance. We thus see that ad hoc contextually determined uses require the context to be such that an unconventional use of a term (e.g., using "ham sandwich" to talk about a customer) plays an individuating role in that particular context.

7.7 More Intrinsic Senses, More Conceptually Determined Ways of Individuating

As discussed earlier in this chapter, "magazine" and "newspaper" differ from "book" in that they can be used to not only to talk about concrete and abstract instances of a given type of reading material but can also be used to talk about the organizations that produce that reading material. Nevertheless, the two differ, in that "newspaper" has an additional sense, which is not available for "magazine." "Newspaper" can appear in sentences like (63), in which it is possible for the newspapers that are read by Sally and Sue to be the same or different concrete instances of the same newspaper (e.g., *The New York Times*) or to be concrete instances of *different* newspapers (e.g., *The New York Times* and *The New York Post*):

(63) After work, Sally will read the newspaper, and so will Sue.

The latter possibility is not available for "magazine." In that case, they must be reading the same (concrete or abstract) magazine. It thus appears that there is a sense in which different newspapers can be thought to be the same, but this sense is not available for "magazine."

That difference is because NEWSPAPER, but not MAGAZINE, individuates the information made shareable by its instances into two formally distinct types of information. First, there is the type of information that newspapers have by virtue of their being newspapers—namely, news; and second, other types of information found in newspapers (e.g., advertising, editorials, crossword puzzles, and other features). The information that newspapers contain by virtue of being newspapers (i.e., news) consists of information about events that happened that readers should know about. That information should be the same in all newspapers. Other information that appears in a newspaper, on the other hand, such as editorials, advertising, and crossword puzzles, is expected to vary across newspapers (publications). This means that if we focus on what newspapers are for, there is a sense in which all newspapers are the same, and thus a sense in which there is only one newspaper. It is this sense that licenses the use of the definite noun phrase "the newspaper" even if different newspapers are read (63). They all make available the news.

DICTIONARY, ENCYCLOPEDIA, CALENDAR, and MAP are like NEWSPAPER in this respect, in that they are all understood to exist for the purpose of making available a specific type of information (definitions, dates, locations) that is not determined by the publication and thus is expected to be the same in all instances of the relevant kind. As such, there is a sense in which the different instances are the same and there is only one instance of the relevant kind, and thus they support weak definite readings of the sort found in (63). On the other hand, BOOK, DIARY, POSTER, PAMPHLET, and CATALOG are all like MAGAZINE; different instances of each of those kinds are expected to contain different information. As such, there is not a sense in which their instances are all the same because they have shared informational content, and thus none of these concepts support weak definite readings.

In sum, concepts like NEWSPAPER license an additional sense that supports weak definite readings such as that in (63). That additional sense results from the concept providing a way of thinking of distinct (publication) instances as being one and the same because they are supposed to share a certain type of informational content that cannot individuate publications (e.g., the news) but does individuate newspapers from other types of reading materials. That possibility is not available for concepts like MAGAZINE, whose abstract instances (publications) are expected to differ in their informational content. Individuation thus enters into explaining another form of systematic polysemy that is intrinsic to certain classes of kind representations. Note that the weak definite interpretation of (63) is available only if we understand Sally and Sue to be reading the newspaper to become informed about the news. That interpretation is not available if they are reading the newspaper only for the editorials or the horoscopes, for example. In weak definite interpretations, "newspaper" is used to talk about just the informational content that is individuated by NEWSPAPER—the news, not the informational content as a whole. The latter can also be referred to via "newspaper"; however, that sense is also available for "magazine," "book," and other similar terms (e.g., "The newspaper/magazine/book/ . . . shocked its readers").

7.8 Individuation and Systematic Polysemy

As Chomsky (2000) has noted, commonsense concepts provide intricate ways of identifying and individuating the things that we think and talk about. In this chapter, I have shown that many patterns of systematic polysemy can be accounted for by the Individuation for Polysemy principle according to which when a representation of type A is required for the individuation of a representation of type B, the name that we use to identify representations of type A and the things they are used to talk about can be used to identify representations of type B and the things that they are used to talk about. We have seen that this principle applies when (1) the B-type representations are intrinsic to the A-type representations, wherein kind representations individuate (abstract and/or concrete) instance-of-kind and subkind-of-kind representations of that kind; (2) both the A-type and B-type representations are kind representations; (3) the A-type representation is of an individual and the B-type representation is of a kind; and (4) both A-type and B-type representations are instance-of-kind representations.

Furthermore, we have seen that when there are systematic conceptual connections between two types of things, but no individuation relation between them, systematic polysemy is not licensed. It is when the way of distinguishing and identifying as one for one kind of thing determines how we distinguish and identify as one the other type of thing that we can use the name of the first to think and talk about the second. This formal relation between conceptual forms licenses systematic polysemy, whereas other conceptual relations, even ones that are lawful and explanatory, do not do so unless they are mediating this individuation relation. As such, it appears that it is when the *perspectives* provided by one class of concepts individuate the perspectives provided by another class of concepts that we can use the term for the former to think and talk about the latter. Although our focus has been on how the properties of conceptual form can help explain the conditions in which we find systematic polysemy, going forward, it should be possible to use different patterns of systematic polysemy to discover and further elaborate the often nonobvious conceptual forms that encode the rich and intricate perspectives provided by our lexically expressible concepts.

8 Metaphorical Perspectives

My lawyer is a shark.
Some ideas are diamonds.
Lies are daggers.

It is remarkable that these sentences are not dismissed as the ramblings of a confused mind. Far from being classified as gibberish, metaphors are ubiquitous in everyday conversation, literary works, and scientific discourse (Camp, 2019; Gibbs, 2008; Lakoff & Turner, 1989). They are also the subject of a great deal of psychological research. The theory of conceptual form provides a new way of thinking about metaphors and their interpretation that differs from the approaches to metaphor comprehension that have dominated psychological research. Because my focus is on what the theory of conceptual form can bring to the study of metaphors, I will consider those approaches only in passing. Readers may consult Holyoak and Stamenkovic (2018), Gibbs (2008), and Glucksberg (2008) and the sources cited therein for discussions of existing approaches and the empirical research that bears on them.

My focus is this chapter is on metaphors that take the form of a categorization statement in which instances of a given kind, or the kind itself, is identified as being a different and incompatible kind of thing (e.g., "My lawyer is a shark"/"Lawyers are sharks"). This is the most studied type of metaphor in psychology. After a brief sketch of the standard approaches to metaphor and a specification of their key limitations, I outline the perspective identification approach to metaphor. I then present some empirical evidence in support of that approach. Finally, I discuss a number of formal constraints on the interpretation of metaphors that come to light when we consider metaphor from the perspective provided by the theory of conceptual form. Those constraints are naturally explained by the conceptual form of kind and instance-of-kind representations, but they are left unexplained by standard approaches to metaphor and conceptual representation.

8.1 Standard Approaches to Conceptual Representation and Metaphor Comprehension

Although ubiquitous, the very existence of metaphors is unexplained, at best, by standard approaches to conceptual representation. Two main approaches have arisen in the study of metaphor comprehension (Glucksberg, 2011; Bowdle & Gentner, 2005; Holyoak & Stamenković, 2018). Here is how these approaches depend on standard approaches to conceptual representation.

8.1.1 Similarity-Based Approaches

As discussed in chapter 1, standard approaches to conceptual representation view categorization as the primary function of concepts and represent categories via summary representations of the properties and causal-explanatory structure that members of the category necessarily or typically share. The concept is understood as categorizing things that meet those conditions, not things that do not meet those conditions. For example, the concept SHARK is used to categorize things that, depending on the approach to concepts that you favor, meet the definition, are similar enough to the prototype, or possess the causal-explanatory structure that causes the category-typical properties of sharks. For obvious reasons, this concept cannot be used to categorize things that belong to other categories, such as the categories represented by DOG, TABLE, or LAWYER. Those concepts have different definitions, prototypes, and causal-explanatory structure from that found in SHARK and thus categorize things in different and incompatible ways. It is for this reason that there is a long tradition of denying that metaphors are anything other than similes in disguise. Sharks cannot be dogs, tables, or lawyers; and lawyers, dogs, and tables cannot be sharks. But sharks can be *like* dogs, tables, and lawyers and vice versa. Metaphorical statements such as (64), which appear to categorize something as something other than what it is, are understood as expressing similes such as (65) despite their form. Because the primary function of concepts is to categorize things, and the form of metaphors such as (64) require us to categorize things in ways that are impossible, we must assume that the form of the metaphor should be ignored. It is reasonable, then, to assume that the speaker who uses a metaphor is alerting the listener to some way in which the topic (e.g., "My lawyer") and the vehicle (e.g., "shark") are similar. Metaphors are really similes according to this view, and theories of metaphor comprehension have been aimed at discovering how we select the appropriate similarities between topics and vehicles via feature-mapping (e.g., Ortony, 1979) or structure-mapping processes (e.g., Bowdle & Gentner, 2005) to comprehend a metaphor:

(64) My lawyer is a shark.

(65) My lawyer is like a shark.

8.1.2 Categorization-Based Approaches

Opposed to the view that metaphors are fundamentally comparisons (similes) is the view that metaphors, at bottom, involve categorization, just as their form suggests. According to this view, interpreting a metaphor requires categorizing the topic (e.g., "my lawyer" in (64)) as belonging to an abstract category implied by the metaphor vehicle that includes the topic and the vehicle (e.g., "shark" in (64)). For example, the topic and vehicle category in (64) can be thought to be members of the abstract category of predatory creatures (Glucksberg, 2011). Furthermore, because the vehicle category is taken to be emblematic of the unnamed abstract category that it and the topic belong to, the unnamed abstract category inherits the vehicle category's name (e.g., "shark"). As such, the vehicle category's name is understood to have dual reference which can be used to talk about the vehicle category (SHARK) as well as the abstract category implied by the metaphor. Given all this, metaphorical statements such as (64) can be understood as involving categorization, albeit it requires us to identify or construct some odd categories that include, for example, lies and daggers, ideas and diamonds, or lawyers and sharks. It also depends on the possibility of referring to those abstract categories via the name of the vehicle category.

8.1.3 Other Approaches

A hybrid of the similarity- and categorization-based approaches has been proposed by Bowdle and Gentner (2005), in which metaphors start their "career" expressing analogy-based structural similarities, but as they become conventionalized, it becomes possible to induce the abstract categories that are central to the categorization view. Holyoak and Stamenkovic (2018) also identify a *conceptual mapping approach* that derives from work in cognitive linguistics; however, it is unclear what psychological mechanisms underlie this approach (or even how coherent some versions of it are; see Murphy 1996). Insofar as it is coherent, it appears to rely on the analogy-based structure mapping and alignment processes central to the similarity-based view of metaphor (Bowdle & Gentner, 2005).

8.2 Do We Form Categories When Understanding Metaphors?

The creation of a category that includes the topic and the vehicle that can be referred to by the name of the vehicle category is crucial to the categorization-based approach to metaphor. In the example that we have been considering, a category includes sharks and my lawyer that can be referred to via "shark." But do we really form a category representation in which "the predicate (metaphor vehicle) refers to a category that includes both the metaphor topic and the metaphor vehicle as exemplar, with the vehicle being a prototypical exemplar of that ad hoc attributive category" (Glucksberg & Keysar, 1990, 13)? A category in which "the topic *lawyers* is included into the abstract category that represents all vicious and predatory things and people" (Haught, 2013, 261)?

The putative category should include my lawyer, sharks, and well-known predators like tigers and lions, which are typically thought to be aggressive, predatory, and vicious. As such, we should be able to make statements about the predators that belong to this abstract category using "shark." But this does not seem to be possible. If the only members of the putative abstract category are the topic and vehicle of a metaphor (66a), the statement sounds like a joke or a clever use of language, as it plays on the literal and metaphorical senses of "shark." If, however, we include the location where *other* members (e.g., tigers and lions) of this abstract category are found (66b), the statement sounds odd and confused rather than merely zeugmatic. This is because neither the literal nor the metaphorical sense of "shark" applies to tigers and lions (or any other kind of predator) that may live in jungles, savannahs, or elsewhere, even though tigers, lions, and other predators have the abstract characteristics that sharks and lawyers are understood as sharing. Similarly, whereas a metaphorical reading of "shark" is licensed when characterizing some lawyers (67a), such a reading is no longer available when other members of the putative abstract category are included (67b):

(66a) Sharks can be found in oceans and law offices.

(66b) #Sharks can be found in oceans, jungles, savannahs, and law offices.

(67a) Some lawyers are sharks.

(67b) #Tigers, lions, and some lawyers are sharks./#Some lawyers, tigers, and lions are sharks.

These observations raise doubts that we form an "abstract category that represents all vicious and predatory things and people," which can be referred to by the name of the vehicle.

It should be noted that there is nothing incoherent about an ad hoc category that includes lawyers, sharks, tigers, lions, and other predators. We can form a category of that sort and refer to it via a description such as "persons and animals that pursue their targets in an aggressive, vicious, and merciless manner," but, as examples (66) and (67) suggest, we cannot refer to such a category via the name of a category that is emblematic of it. Similar statements using a description of the ad hoc category, on the other hand, sound fine (68):

(68a) Persons and animals that pursue their targets in an aggressive, vicious, and merciless manner can be found in oceans, jungles, savannahs, and law offices.

(68b) Tigers, lions, and some lawyers are creatures that pursue their targets in an aggressive, vicious, and merciless manner.

If understanding the metaphor involves forming an ad hoc abstract category that is referred to via the vehicle category's name, it appears that the category cannot include nonmetaphorical predators such as tigers and lions, as they cannot be referred to via the vehicle's name. Perhaps the ad hoc category contains only the vehicle category and things that are fundamentally different kinds of things from the vehicle category that share key abstract characteristics with the vehicle category. In the example that we have been considering, the reformulated category would contain sharks and things that are of a fundamentally different sort than sharks but share characteristics with sharks such as being vicious and aggressive. This is a very odd category. It contains sharks and some lawyers, bankers, and poker players because they are vicious, aggressive, and predatory, but it excludes tigers and lions, despite the fact that they also display those characteristics. Categories usually contain things that are thought to be the same sort of thing in some respect. Excluding items of a similar kind that display the shared characteristics (e.g., being aggressive, vicious, or predatory) from the category while including those that are of a fundamentally different sort is bizarre. Furthermore, it has no motivation unless the category membership conditions also include the fact that all the members can be spoken about via the name of the vehicle. However, adding the reference condition renders the categorization approach stipulative and unexplanatory.

It is important to note that the metaphorical sense of "shark" *can* be used to talk about multiple different kinds of predators that may be thought to be members of the ad hoc category that is generated if they are all fundamentally different sorts of things from the vehicle category. For example, (69) is a perfectly fine metaphorical statement that identifies a number of different types of people as sharks:

(69) Some lawyers, bankers, and poker players are sharks.

(70) Sharks can be found in oceans, law offices, banks, and casinos.

On the other hand, (70) is zeugmatic in the same way as (66a), as it plays on the literal and metaphorical senses of "shark." The zeugma suggests that whereas "shark" has a literal and a metaphorical sense, the metaphorical sense cannot be used to talk about sharks as a member of the category that includes some instances of LAWYERS, BANKERS, and POKER PLAYERS. This is problematic if "the predicate (metaphor vehicle) refers to a category that includes both the metaphor topic and the metaphor vehicle as exemplar, with the vehicle being a prototypical exemplar of that ad hoc attributive category" (Glucksberg & Keysar, 1990, 13). It appears that the metaphorical sense of "shark" can be used to refer to the ad hoc category that is generated, but the generated category does not include the metaphor vehicle itself as an exemplar, as reference to the vehicle category seems to require the literal sense of "shark," and thus the zeugma of (66a) and (70). This suggests that the metaphoric use of "shark" is not licensed by SHARK being a prototypical member of the ad hoc abstract category that is generated by the metaphor. Some cases of dual reference, such as the use of "Kleenex" to refer to a specific brand of facial tissue as well as facial tissues generally, are based on the general category receiving its name from the specific brand, as the brand is understood to be a prototypical member of and emblematic of the general category. Crucially, in such cases, the brand is an instance of the general category and thus we can say things like (71):

(71) One kind of kleenex is called "Kleenex," another is called "Scottie."

But it appears that SHARK is not a member of the category of metaphorical sharks. This means that there must be some other mechanism for licensing the metaphorical sense of "shark," as well as vehicle category names more generally.

There is one type of case, however, in which the dual-reference mechanism posited by the categorization approach clearly works. These are cases in which a novel category is generated on the basis of an individual exemplar and its proper name. Glucksberg (2011) discusses the example of "Enron" being used not only to refer to the individual company that failed due to its scandalous accounting practices, but also to any company that fails for similar reasons, as in "Will Worldcom be the next Enron?" A similar example involving the use of "Florida" to refer to any place involving an election controversy concerning the proper counting of ballots is discussed by Glucksberg and Haught (2006) and Glucksberg (2011), who cite an election monitoring official as saying, "Unless we do something now, Florida is headed toward being the next Florida*" (Glucksberg, 2011, 15). Glucksberg understands the first use of "Florida" as a literal use and the second use (with the asterisk) to be metaphorical—"the literal 'Florida' can only *be* a metaphorical 'Florida' because the 'next Florida' does not yet exist" (Glucksberg, 2011, 15). In both the Enron example and the Florida example, a novel category is clearly generated on the basis of an individual with some distinctive characteristics, and the category includes all things that have the relevant distinctive characteristics. The individual from which the category is generated is clearly included in the category and is understood to be emblematic of the category.

Finally, the category inherits its name from the individual instance from which the category is generated. This is exactly how the categorization approach to metaphor is proposed to work. The problem is that it is unclear that there is anything metaphorical going on in the Enron and Florida examples. In these examples, the name of the individual is a proper noun and the name for the category is a common noun that shares the phonological form of the proper name. It appears that there are proper nouns that literally refer to an individual company or place and a common noun that literally refers to companies or places of a certain sort. The category name does not seem to have both a literal and metaphorical sense, just a literal sense. This is entirely parallel with the Kleenex type cases discussed previously. This also contrasts with cases like "shark," in which a single common noun has both a literal sense and a metaphorical sense.

It is true, as Glucksberg notes, that a simile that compares the literal Florida to the next Florida (e.g., "#Florida is like the next Florida") is infelicitous because the next Florida does not yet exist, but his conclusion, that the second use of "Florida" in the quote is metaphorical, does not follow. The same infelicity appears in "#Florida is like the next state to have an electoral ballot-counting controversy" for exactly the same reason, even though there is no possibility of metaphorical interpretation in this example. All kinds have both actually existing instances and potentially existing instances (including a potentially existing instance that will become actual next). The fact that there are instances of a kind that exist only potentially that do not yet exist actually is a formal characteristic of kind representations, not metaphorical representations (see chapter 4).

All of this suggests that, as intuitively appealing as the categorization approach is, there appear to be difficulties in spelling out exactly how it is supposed to work. The comparison approach, which simply assumes that metaphors are similes in disguise, is not a viable theory of metaphor, as it leaves the form of metaphors unexplained and because metaphors and similes can differ in their meanings (Glucksberg, 2011; Haught, 2013). Of course, this does not rule out the possibility that comparison processes are a crucial part of metaphor comprehension, nevertheless (see section 8.8).

8.3 Metaphor from the Perspective of Conceptual Form

According to the theory of conceptual form, lexically expressible kind concepts such as DOG, TABLE, LAWYER, and SHARK provide perspectives for thinking and talking about different kinds and their instances. Those perspectives are mental lenses for thinking and talking. Furthermore, we have argued that the metaphor of a "mental lens" can be cashed out in terms of a concept's formal structure and the instructions that it provides for thinking about the contentful elements of a concept. Each concept possesses a formal structure that provides instructions in terms of a set of principles for how to think and talk about the kind, its instances, and the properties that characterize and distinguish the kind from other kinds. For example, the concept SHARK provides a set of instructions for how to think and talk about something if we think of it as an instance of the kind "shark," as well as how to think and talk about the kind itself. That same set of instructions, and thus the perspectives that they encode, can also be used to pretend that something that is not a shark, such as a stapler, is a shark. Concepts, as per this view, do not specify how to categorize things. Instead, they specify how to think about something *given* that you have reason to think of it as being an instance of the kind represented by the concept. For example, when a child says something like (72), they are likely to be pretending that the stapler is a shark, and the concept SHARK will guide their pretense. For example, the child may make the stapler swim, prey on smaller fish, or attack a bather at the beach. The child may also use "shark" to talk about the stapler while engaging in the pretense:

(72) That stapler is a shark.

To engage in this pretense, the child need not construct an ad hoc category that includes the stapler and sharks in order for knowledge about sharks to guide how the child thinks and talks about, or acts on, the stapler. And the child should certainly not modify the concept SHARK to include things that have the properties of the stapler in it (Leslie, 1987)! Instead, the concept SHARK is used to provide instructions about how to think and talk about the stapler while the child is pretending that it is a shark. The concept does not categorize the stapler. Instead, it provides the perspective from which to think and talk about the thing that we are pretending is a shark. All of this is possible because lexically expressible kind concepts such as SHARK provide perspectives for thinking and talking about things that we take to be instances of that kind or the kind itself. As such, kind concepts do not categorize but presuppose categorization. In this, the present view of lexically expressible concepts departs from standard views of conceptual representation.

Pretend scenarios are one context in which the concept for one kind of thing (e.g., SHARK) and its name is used to think and talk about some other kind of thing (e.g., STAPLER). As we saw in chapter 7, systematic polysemy presents another context in which the name and concept for one kind of thing can enter into thinking and talking about another kind of thing. For example, "magazine" can be used to talk about a certain type of reading material (e.g., "The magazine on the coffee table shocked many of its readers"), but it can also be used to talk about the organization that produces that reading material (e.g., "The magazine fired its editor"). In neither of these cases does the use of a word used to talk about one kind of thing to talk about a very different kind of thing require forming an ad hoc category. In a pretend scenario such as (72), the properties of the subject are simply ignored, and the predicate determines the way that we think and talk about the subject. In the systematic polysemy case, kind concepts of a given class (e.g., kinds of reading materials) enter into determining what counts as one instance of kind concepts of another class (e.g., the organizations that produce those reading materials), and the name for the former kinds are used to talk about the latter kinds (see chapter 7).

Metaphors present us with a situation in which the vehicle concept is used to think and talk about the topic. In this case, we are not pretending that the topic is an instance of the vehicle category. Instead, metaphors ask the listener to think of the topic from the perspective provided by the vehicle category. In the example that we have been considering, a metaphorical interpretation requires us to think about my lawyer and his properties from the perspective provided by SHARK—that is, to attribute the kind identity or nature of a shark to my lawyer and to use some of the instructions provided by the conceptual form of SHARK to think and talk about my lawyer and his properties. So, for example, in calling my lawyer a shark, the metaphor suggests that being vicious is part of his nature since being vicious is represented as an aspect of being a shark. Although many lawyers may be vicious, we do not think that being vicious is an aspect of being a lawyer. There is only a statistical connection between being a lawyer and being vicious, but there is a principled connection between being a shark and being vicious. Consequently, in identifying my lawyer as a shark, the metaphor suggests that when my lawyer acts in a vicious manner, he is acting in accord with his nature. It is not out of character for him to act in a vicious manner if we think of him from the perspective provided by SHARK, and we can explain his being vicious by stating that he is a shark. If we simply identify him as a lawyer, on the other hand, we do not have a reason to think that his vicious actions are due to his nature, as we do not think that there is a principled connection between being a lawyer and being vicious.

Using the formal structure of SHARK to think about my lawyer thus leads us to think about his actions differently than if we use LAWYER to think about those same actions.

In sum, because commonsense concepts are understood as providing perspectives for thinking and talking about things, rather than specifying what needs to be true of things in order for them to be categorized as belonging to a given category, it is possible to understand metaphors as identifying the perspective from which the speaker would like the listener to think and talk about the topic. We may call this the *perspective identification* view of metaphor.

8.4 Fleshing Out the Perspective Identification View of Metaphor

Given that the vehicle category provides the perspective from which the metaphor asks the listener to think and talk about the topic, one source of constraint on the interpretation of a metaphor comes from the formal structure of the vehicle category. The formal structure of a basic-level category such as SHARK identifies (1) the sort of thing that something is understood to be, (2) what counts as one instance of the kind, and (3) the respect in which instances are expected to remain the same throughout their existence. Such sortal principles are part of the perspective that basic-level kind concepts such as SHARK provide, but they are not relevant to the interpretation of "shark" when it is the vehicle of a metaphor, although they are relevant to pretend scenarios that use SHARK. In a metaphor, the topic concept provides the sortal principles that identify what we are talking about. In addition to the sortal principles provided by the conceptual form of a kind concept, the conceptual form specifies the character of the kind that characterizes and distinguishes the kind from other kinds in terms of the properties that have a principled connection to the kind and are understood to be aspects of being that kind of thing. It is this component of the perspective provided by a vehicle concept that is relevant to metaphor interpretation. In a pretend scenario, the properties that have a principled connection to the kind are simply assumed to be true of the subject. Metaphor interpretation is more selective.

In a metaphor, the vehicle concept attributes a property or a set of properties to the topic but does not fully characterize the topic as it does in a pretend context. To get clearer on how metaphors work, it is useful to consider how metaphorical attributions are similar to, and different from, literal attributions. Both literal and metaphorical attributions are constrained by whether the topic is a kind (e.g., "Lawyers are sharks") or an instance or set of instances of a kind (e.g., "My/some lawyers are sharks"). If the topic is a kind, then both literal and metaphorical attributions are constrained to those properties that can characterize the kind, which are obviously a subset of the properties that can be attributed to one or more instances of the kind, but not the kind in general. For example, the literal statement "Some ideas are clear and valuable" is true, and the metaphor "Some ideas are diamonds" sounds like an apt way of conveying that thought, but "Ideas are clear and valuable" is false, as there are many ideas that may be neither, consequently, the metaphor "Ideas are diamonds" sounds odd and untrue.

It is important to note, however, that the literal and the metaphorical attribution of a property need not match in the *manner* in which they attribute a given property to a topic. For example, as noted previously, although we may literally characterize lawyers as vicious, aggressive, and ruthless, such characterizations posit a statistical connection between LAWYER and the attributes. As such, we do not think of these properties as aspects of being a lawyer, and we do not think that we can explain their presence in a lawyer by citing the fact that the person is a lawyer, nor do we think that lawyers are *supposed to* possess these properties (see chapter 6). However, in the metaphor, lawyers are attributed the nature of a shark and thus are understood to have a principled connection to being vicious, aggressive, and ruthless (73):

(73) Lawyers are sharks. It is in their nature to be vicious, aggressive, and ruthless.

Although it needs to be systematically investigated, it is possible that this difference in the manner in which the attributed property is connected to the topic in literal and metaphorical attributions may explain the intuition that metaphoric attributions are stronger than literal attributions or similes.

Of course, there are metaphors in which the attributed property has a principled connection to both the topic and the vehicle. For example, in the metaphor "Beavers are lumberjacks," the property CUT DOWN TREES has a principled connection to both LUMBERJACK and BEAVER. In such cases, even though the vehicle does not provide a different way of thinking of how the attributed property is connected to the topic, the metaphor may highlight the fact that the property is understood to be part of the topic's nature. The form of the literal attribution "Beavers cut down trees" does not convey the idea that it is in the nature of beavers to cut down trees. On the other hand, the metaphor "Beavers are lumberjacks" attributes the conceptual form of LUMBERJACK to them. As such, it conveys the idea that cutting down trees is in their nature and thus implies an explanation for why they do it, that they are good at it, and that they do it regularly.

As the "Lawyers are sharks" and "Beavers are lumberjacks" metaphors illustrate, the vehicle can specify how to think about properties that the topic *actually* has for both properties that have a statistical connection to the topic and those that have a principled connection to the topic. It also is possible for metaphors to attribute properties that are not *actually* true of the topic but are *potentially* true of it. For example, it is unlikely that people think that lawyers can generally be characterized as merciless—even those who may be willing to characterize lawyers as vicious and aggressive may reject the idea that they can also be generally characterized as merciless, even if some may be so characterized. If so, "Lawyers are merciless" would be understood to be false. Nevertheless, the metaphor "Lawyers are sharks" can be interpreted as attributing the property of being merciless (among others) to lawyers generally (Glucksberg, 2008). Such examples show that a metaphor can attribute a characteristic that is only potentially true of a topic, in addition to specifying how to think of properties that are actually true of the topic.

The vehicle concept cannot attribute properties to the topic that are neither actually or potentially true of the topic. This explains why properties like HAS FINS, LEATHERY SKIN, and SHARP TEETH are not attributed to lawyers in the metaphor that we have been considering. Those properties are neither actually or potentially true of lawyers. Lawyers are not constituted such that under the right conditions, they would grow the fins, leathery skin, or sharp teeth of a shark.

Literal attributions can involve a single property or a combination of properties. For example, lawyers may be characterized as vicious, aggressive, and ruthless. Any of those attributions could be expressed via the metaphor "Lawyers are sharks," but the metaphor can also be used to attribute all the properties at once. The literal attributions do not confer any nonaccidental unity to the different properties attributed. So, "Lawyers are vicious, aggressive, and ruthless" is no different from "Lawyers are well paid, argumentative, and easy to find" in that in both cases, multiple properties are attributed and no unity among the properties is implied. In contrast, metaphors specify the conceptual form of a kind for thinking about the properties of the topic, and thus they specify an intrinsic unity among a set of properties that are aspects of being that kind of thing.

8.5 Emergent Properties in Metaphor Comprehension

It has often been noted that in understanding a metaphor, participants list properties that the metaphor attributes to the topic that participants do not list as properties of either the topic or the vehicle outside a metaphor. The properties that are listed only in the context of a metaphor are called *emergent properties*. The perspective identification view of metaphor can explain the generation of emergent properties.

Thus far, we have been specifying the ways in which the conceptual form of the vehicle concept guides the ways in which we think about properties of the topic. But to do so, the conceptual form of the vehicle considers only certain aspects of the vehicle concept as providing instructions for how to think about the topic. As such, the metaphor also provides a different perspective on the vehicle concept by highlighting those properties of the vehicle concept that are relevant to the topic. In such a context, properties that may not characterize the vehicle concept generally, such as MERCILESS, may become prominent in a context in which aggressive and vicious are highlighted because they correspond to properties of the topic. This illustrates how emergent properties become available in metaphorical contexts given the perspective identification view of metaphor.

8.6 Some Evidence for the Perspective Identification View of Metaphor

A number of experiments that we are currently conducting in my lab provide some initial evidence for the perspective identification view of metaphor. One set of experiments tested the hypothesis that the property attributed to a topic in a metaphor must have a principled connection to the vehicle category, but it need not have a principled connection to the topic category. Shamauri Rivera and I investigated this hypothesis with a set of metaphors taken from Jones and Estes (2006). We included all the highly apt metaphors that had count nouns for both the topic and vehicle from that study (you'll see why we limited our items in this way in a moment). We also added the classic "Lawyers are sharks" metaphor so that we had an even number of items. As all the items that we chose were highly apt metaphors, it was clear what property they were attributing to the topic. Given the hypothesis that the attributed property has a principled connection to the vehicle category but need not have a principled connection to the topic category, we predicted that statements that made an explicit formal explanatory connection between the vehicle and the attributed property (e.g., "Sharks, by virtue of being sharks, are vicious") would be judged to be true to a greater extent than those that did the same for the topic (e.g., "Lawyers, by virtue of being lawyers, are vicious"). This is exactly what we found with participants judging the statements involving the vehicle to be true to a greater extent (mean rating = 5.8) than the statements involving the topic (mean rating = 4.3).

Principled connections also support normative expectations, so a second experiment tested whether there is a normative expectation concerning the attributed property vis-à-vis the vehicle category to a greater extent than for the topic category. Participants judged statements that expressed a normative expectation for the attributed property in the vehicle category (e.g., "Sharks are supposed to be vicious") and the topic category (e.g., "Lawyers are supposed to be vicious"). As predicted, participants judged the statements involving the vehicle to be true to a greater extent (mean rating = 6.1) than those involving the topic (mean rating = 4.6).

Finally, properties that have a principled connection to a kind are expected to generally be present in instances of that kind. Accordingly, we predicted that participants would estimate the prevalence of the attributed property to be higher in the vehicle category than the topic category. This prediction was confirmed with participants estimating the attributed property to be true of 86.1 percent of instances of the vehicle category and 57.3 percent of instances of the topic category.

Together, these three experiments provide evidence for the perspective identification theory of metaphor by showing that the property attributed to the topic in a metaphor is more likely to display the signatures of a principled connection to the vehicle category than to the topic category.

A different test of the hypothesis comes from an experiment with novel metaphors that Cameron Reicher and I conducted. In that experiment, we presented novel metaphors to participants and asked them to rate how likely it was that the metaphor was being used to attribute a property to the topic that either had a principled connection to the vehicle or a statistical connection to the vehicle. For example, given the novel metaphor "My shark is a lawyer," participants were asked to rate how likely it was that the speaker used this metaphor to convey the idea that my shark is argumentative, a property that has a principled connection to the vehicle LAWYER, or the idea that my shark is well paid, a property that has a statistical connection to the vehicle LAWYER. The perspective identification approach to metaphor predicts that participants should think it is more likely that the metaphor was used to attribute the property of being argumentative than the property of being well paid. Of course, it may be that as implausible as both ideas are, the properties that have a principled connection to the vehicle are more plausibly true of the topic than the properties that had a statistical connection to the vehicle. If the participants thought that, then the predicted difference would arise, but for a different reason than predicted. As such, we also asked participants to tell us how plausible it was that the topic had each of these properties. We found that participants did in fact think that it was slightly more plausible for the topic to have the property that had a principled connection to the vehicle (mean rating = 3.2) than the property that had a statistical connection to the vehicle (mean rating = 2.8). However, that small difference could not account for the large difference found between the two types of properties in the interpretation of the metaphors. Participants thought the metaphor was much more likely to attribute the property that had a principled connection to the vehicle (mean rating = 4.0) than the property that had a statistical connection to the vehicle (mean rating = 2.6). These results provide further evidence that metaphor interpretation involves attributing a property that has a principled connection to the vehicle to the topic.

Finally, an experiment carried out by Dominick Ferrara and I provided some evidence for the idea that in interpreting a metaphor, the conceptual form, and thus the nature of the vehicle category, are attributed to the topic. In the experiment, on some trials, participants saw a passage such as "My lawyer is a shark. It is in his nature to be vicious." On other trials, they saw passages such as "My lawyer is like a shark. It is in his nature to be vicious." On all trials, participants were asked to rate how natural the second sentence sounded in the context of the first sentence. It was predicted that the sentence describing the attributed property as being part of the topic's nature would sound more natural after a metaphor than after a simile. This is because, by hypothesis, interpreting a metaphor involves thinking of the topic from the perspective provided by the conceptual form of the vehicle, and thus to think of the attributed property as a part of the topic's nature. Interpreting a simile, on the other hand, does not require identifying a similarity that is considered to be part of either the topic's or the vehicle's nature or attributing the vehicle's nature to the topic. The results of the experiment confirmed this difference in the way that metaphors and similes are interpreted, with the sentence identifying the attributed property as part of the topic's nature being rated as more natural when it followed a metaphor (mean rating = 3.3) than when the same sentence followed the corresponding simile (mean rating = 2.5).

These experiments provide initial evidence for the perspective identification view of metaphor that is rooted in the theory of conceptual form. We are in the process of designing and running more experiments that we expect will test and refine the theory further. Next, I discuss some further formal constraints on the interpretation of metaphors that are naturally accounted for by the perspective identification view of metaphor but are unlikely to be easily addressed by existing approaches.

8.7 More Formal Constraints on the Interpretation of Metaphors

8.7.1 Metaphors and the Paradox of Unlike Things Compared

Similes allow us to compare like and unlike things. "Whales are like sharks" and "Lawyers are like sharks" are both perfectly good similes. On the other hand, whereas "Lawyers are sharks" is a perfectly good metaphor, "Whales are sharks" is not only false but sounds incoherent and has no metaphorical reading. Glucksberg (2011, 7) describes this state of affairs in the following manner: "That two unlike things can be expressed as an identity relation while two like things cannot is a paradox—the paradox of two unlike things compared . . . What is it about metaphorical categorical assertions that afford paraphrasing as similes (and vice versa) while literal categorical assertions do not?"[1] The theory of conceptual form provides an account of this paradox.

Recall that according to the theory of conceptual form, there are no bare kind representations. Every kind representation is a kind representation from a formally specified class of kind concepts. For example, DOG and CAT are kind concepts from the class of animate kind concepts. Each kind concept within the class has the same sort of formal structure, but each represents a different animate kind, distinguished by its unique character and an index (usually the kind's name) that represents the kind's actual existence (see chapter 5). Because the kind representations within a class share their formal structure, their distinctness and incompatibility are immediately evident. Furthermore, one concept cannot provide a different perspective for thinking about the other concept because they provide the same sort of perspective.

On the other hand, kind representations such as LAWYER and SHARK come from different classes and thus differ in their formal structures, characters, and actuality indices. As such, while the distinctness of the kinds is immediately apparent, their incompatibility is not immediately evident. We cannot assume incompatibility simply on the basis of distinct formal structures because we can often think about kinds from more than one perspective, using concepts with different formal structures. For example, DOG and DOMESTIC ANIMAL provide different perspectives that are compatible with one another—dogs are domestic animals. The concept DOG provides a perspective for thinking about dogs as a kind of animate being, while DOMESTIC ANIMAL highlights their role as companions for humans, typical modes of interaction between humans and them, and other elements. Such characteristics are understood to be aspects of the character of DOMESTIC ANIMAL, but they are not understood to be aspects of the character of DOG. In other cases, for example, with the concepts LAWYER and SHARK, the perspectives provided by the two concepts are not compatible—lawyers are not sharks. But because LAWYER and SHARK provide different sorts of perspectives, a metaphorical interpretation is possible. As discussed previously, in such cases the conceptual form of the vehicle partially determines how we think about the topic's character.

The theory of conceptual form thus explains the paradox of unlike things compared. Two like things cannot be expressed as an identity relation (e.g., "#Dogs are cats") because they are distinct and incompatible things of the same sort. Furthermore, they possess the same formal structure, so there is no possibility of one concept being said to be the other one in a metaphorical sense, wherein one concept provides a different perspective from which to think about the other. On the other hand, an identity can be expressed between two unlike things despite the fact the things are not the same, literally because the unlike things have different formal structures, which allows for the possibility of thinking that there is a metaphorical sense in which they are the same.

8.7.2 The Instance-of-Kind Exception

As noted earlier, we cannot make categorical statements concerning similar kinds (e.g., "#Whales are sharks") that can be understood as metaphors. We *can*, however, make metaphorical statements involving those same kinds if the topic is an instance or some instances of the topic kind rather than the kind itself. So, for example, we can say things like "Some whales are sharks" to express the idea that some whales are vicious, aggressive, and predatory. A more familiar metaphor of this sort may be (74), to express the idea that my cat is affectionate and seeks my attention:

(74) My cat is a dog.

Similarly, the metaphor "My dog is a cat" can express the idea that my dog is aloof and independent. Why is a metaphorical reading possible in such cases? The theory of conceptual form provides an answer.

Instance-of-kind representations have a formal structure that provides the means for thinking about things insofar that they are instances of a given kind. As such, they are understood to have certain properties by virtue of their being the kinds of things that they are, but they also have accidental properties that are not determined by the kind. Accidental properties are true of instances for reasons other than the instances being the kind of thing that they are. It is even possible for instances to have properties that contradict the properties that instances of that kind are expected to have in virtue of their being the kind of thing that they are. For example, there can be three-legged dogs even though dogs, by virtue of being dogs, are four-legged. Of course, the three-leggedness of particular instances is understood to be for reasons other than their being dogs. Because properties of an instance of a given kind can be determined by something other than the kind of thing that it is, it allows for the possibility that an instance may be characterized by a perspective provided by a kind other than the kind of thing that it literally is.

Kind representations, on the other hand, represent the nonaccidental intrinsic qualitative and quantitative unity of a kind and its instances. Neither the kind nor its character is understood to be determined by anything other than the kind itself; as such, there is no possibility of another kind being relevant for characterizing the kind unless it is a kind with a different sort of formal structure, in which case the other kind can provide a different perspective for thinking about the kind and its properties.

The difference in the conceptual forms of kind and instance-of-kind representations thus provides an explanation for why kind concepts from the same formally distinguished class cannot be used to generate a metaphor, while also allowing that instances of a kind may be metaphorically characterized by another kind from the same formally distinguished class of kind representations.

8.7.3 The Exceptions Are Exceptional

Importantly, the theory of conceptual form also suggests a difference in how metaphors involving like kinds are interpreted in comparison to standard metaphors that involve fundamentally different kinds of things, such as "Lawyers are sharks." Interpreting metaphors such as "Some dogs are cats," which has a topic that refers to instances of the topic and a vehicle which is from the same formal class of kinds as the topic, requires understanding the instances as exceptional. The metaphorical nature of the instances is understood to be in opposition to the nature that they are supposed to have. Evidence for this comes from the fact that sentences such as (75) sound natural. That opposition between the metaphorical nature of the topic and the topic's nature seems not to exist when the topic and the vehicle come from formally distinct classes (76):

(75) Some dogs have the nature of a cat even though they are dogs.

(76) #Some lawyers have the nature of a shark even though they are lawyers.

In both types of cases, the topic cannot literally be the vehicle, and they afford metaphorical interpretations; however, when the topic and vehicle come from the same formal class of kinds, there is understood to be an opposition between the topic's true nature and its metaphorical nature, but no such opposition is understood to hold when the topic and the vehicle come from formally distinct classes of kinds.

8.7.4 A Limit on Exceptions

Further evidence for the exceptional interpretation given metaphors such as (75) is that they are limited to cases in which it is reasonable to assume that the topic has the metaphorical nature by accident—where the number or proportion of instances that have that nature is small (77). On the other hand, if a majority or high proportion of instances are said to have the metaphorical nature, it seems that such a situation could not arise by accident, so the metaphorical interpretation is no longer possible (78):

(77) A/some/many dogs are cats.

(78) #Most/#almost all/#all dogs are cats.

(77a) A/some/many lawyers are sharks.

(78a) Most/almost all/all lawyers are sharks.

If the attributed nature is not accidental, then it cannot be merely characterizing some instances of the kind; rather, it must be characterizing the kind itself, but that leads to a contradiction that cannot be avoided via a metaphorical sense for the reasons discussed earlier in this chapter. This restriction on the proportion of instances that the topic can refer to is not found if the topic and vehicle come from different classes of kind concepts (78a).

We thus see an intricate set of formal constraints on when metaphors are possible and how they must be interpreted. Metaphors are not possible when the topic and the vehicle refer to kinds from the same class of kind concepts, but they are possible if the topic kind and the vehicle kind come from formally distinct classes of kind concepts. If, however, the topic refers to instances of the topic kind rather than the kind, the topic and the vehicle can come from either the same or different class of kind concepts, but they are interpreted differently depending on whether they come from the same or different classes. If the topic and vehicle are from the same class of kind concepts, the instances of the kind that are the topic must be understood to be exceptions, and thus to have characteristics that stand in opposition to the nature specified by the topic's kind. On the other hand, if the topic and vehicle are from different kind classes, the instances that are referred to by the topic are not thought of as exceptions that are in opposition to the nature specified by the topic kind.

Finally, in metaphors in which the topic and the vehicle come from the same class of kind concepts, the topic can refer to instances of the topic kind so long as there are not too many of them, and thus it is plausible to think that the instances possess characteristics that are in opposition to the topic kind's nature by accident; otherwise, the metaphor is not possible. No such limit on the proportion of instances that can be the topic of a metaphor exists if the topic and vehicle come from different classes of kind concepts. This intricate pattern of formal constraints on the use of metaphors is naturally explained by the conceptual form of kind and instance-of-kind representations and the perspective identification approach to metaphor. It is unlikely that it can be explained by standard approaches to conceptual representation and metaphor.

8.7.5 Diverging Interpretations for Metaphors and Similes

Formal constraints on metaphor interpretation are also evident if we consider an example that Glucksberg (2011) and Haught (2013) use to illustrate how the meanings conveyed by metaphors and similes can diverge. They point out that the simile in (79) directs attention to properties of old sharks that are shared by my lawyer, and these are likely to be largely negative things like being weak and slow. In contrast, the metaphor in (80) suggests that my lawyer is not only a shark, but also that he has been so for a long time and thus that he is an experienced shark—a positive thing. This difference in interpretation is understood to be due to "old" modifying the literal and metaphorical senses of "shark" according to the categorization view held by Glucksberg and Haught.

According to the present view, "old" is interpreted as modifying the vehicle in the simile and characterizing the topic in the metaphor. Evidence that "old" is characterizing the topic in the metaphor comes from considering (79′) and (80′), in which the topic and vehicle refer to kinds instead of instances and we don't get the same contrast between the simile and the metaphor because "old" can no longer be understood as characterizing the topic and must be understood as characterizing the vehicle (as it must in the simile). Consequently, both the simile and the metaphor have the same negative interpretation. The reason for this shift in interpretation is because "old" cannot be understood as characterizing lawyers generally (81). Thus, the availability of the positive interpretation depends on the possibility of interpreting "old" as characterizing the topic rather than its ability to modify an abstract sense of the vehicle:

(79) My lawyer is like an old shark.

(80) My lawyer is an old shark.

(79′) Lawyers are like old sharks.

(80′) Lawyers are old sharks.

(81) #Lawyers are old.

Haught (2013) and Glucksberg (2011) show that picking modifiers such as "well-paid," which can modify the topic but not the vehicle, results in metaphors (82) that do not have a simile equivalent (83):

(82) My lawyer is a well-paid shark.

(83) #My lawyer is like a well-paid shark.

In the present account, "well-paid" characterizes the topic, so (82) is understood to be about lawyers who are sharks and are well paid, but the corresponding simile is infelicitous because similes require the modifier to be interpreted as characterizing the vehicle, and sharks cannot be well paid. As such, the perspective identification approach to metaphor can capture when and why metaphors and similes can sometimes diverge in their interpretations.

8.8 Metaphors Are Lenses

Metaphors are lenses. They provide new perspectives for thinking and talking about kinds and instances of kinds. Of course, according to the theory of conceptual form, kind and instance-of kind representations are already perspectival. Thus, DOG, SOME CATS, MY LAWYER, and SHARK all provide perspectives for thinking and talking about different kinds and instances of kinds. They do so because they possess formal structures that provide instructions for how to think that the contentful elements of the concepts are related to the kinds and their instances. Given this, metaphors involve using the perspective provided by one concept (the vehicle) to think about another concept (the topic). As we saw earlier, this can get intricate, but the existence of metaphors and how they may be interpreted is simply a matter of specifying the relevant formal structures and their combination, but there is nothing mysterious involved. On the other hand, if we start with traditional theories of concepts, then we are starting with intrinsically nonperspectival concepts, and it is unclear how perspectives, which are needed for understanding metaphors, arise from the combination of nonperspectival concepts. Insofar as theories of metaphor used standard approaches to conceptual representation, the formal dimension of commonsense concepts remained unavailable for guiding research on metaphor.

In this chapter, I have sketched how the theory of conceptual form allows us to formulate the perspective identification approach to metaphor, along with some formal constraints on the interpretation and use of metaphors. The theory of conceptual form places important formal constraints on the interpretation and use of metaphors of the form "A is B." It does not, and cannot, predict the content of metaphors. It places formal constraints on what the content of metaphors of this form can be, but it does not determine it.

Although this chapter has focused on the role of conceptual form in metaphor use, it is likely that feature matching and structure matching processes help align the conceptual form of the vehicle with the relevant properties of the topic so that the conceptual form of the vehicle can be used to think and talk about the topic. As Camp (2019) notes, the matching process required for interpreting metaphors is looser than the systematic structure mapping required for analogy. Camp suggests that in interpreting metaphors, we often match individual features as well as matching abstract structural relations, and also allow unsystematic relational matches. This allows metaphors to convey qualitative similarities in addition to abstract structural similarities, and also to potentially suggest multiple abstract similarities simultaneously. All of this makes metaphors less precise than analogies in certain ways, but also richer.

Finally, in addition to sketching out the perspective identification approach to metaphor and identifying a set of formal constraints on our use of metaphors, this chapter presented some preliminary empirical evidence in support of the perspective identification theory of metaphor. We are currently running experiments to provide further tests of this theory. Perhaps the seeds that we have planted here will blossom into a fruitful theory of the formal constraints that the theory of conceptual form places on the generation and interpretation of metaphors.

IV Acquisition

9 Acquiring Perspectives

Lexically expressible kind concepts such as DOG, TABLE, and SAND contain not only a representation of the properties and causal-explanatory structure shared by members of those categories, but also a "mental lens," or perspective in terms of which those properties are understood to characterize the kind and its instances. That perspective is encoded in the concept's formal structure, which provides instructions for how to think and talk about the contentful properties of the kind and its instances. Any theory of concept formation or concept representation must account not only for how a concept's content is acquired, but also how the formal structure that encodes the perspectives provided by the concept is acquired. Previous research on conceptual representation and development has not considered the formal structure of concepts or how it may be acquired.

In this chapter, I outline some of the challenges that the acquisition of the conceptual form of kind concepts presents, as well as one potential solution for how we can acquire the conceptual form of concepts. Some of the acquisition challenges were briefly considered in chapter 2; however, that discussion was in very general terms because we did not have an explicit characterization of the conceptual form of kind concepts at that point. In this chapter, I consider the acquisition challenges presented by the conceptual form of kind concepts outlined in chapters 4 and 5 (section 9.1). After laying out the challenges, I will consider two general approaches that one can take to try to meet those challenges (section 9.2). One approach, which I will call the *emergent perspectives* approach, assumes no innate structure or knowledge concerning conceptual form. The other approach, which I will call the *perspective identification* approach, assumes a fair amount of innate structure. But what I want to emphasize at the outset is that even if we take the latter approach, we will still have to explain how children acquire specific conceptual form knowledge. In section 9.3, I outline a solution to the acquisition problem raised by the perspective identification approach. The solution is a bootstrapping solution that takes advantage of systematic and nonarbitrary relations between the representations made available by core knowledge systems and the perspectives provided by certain classes of commonsense concepts. Section 9.4 briefly discusses evidence about which perspectives may be available to infants at the start of language learning. Finally, in section 9.5, I discuss how the form and content of core knowledge systems differ from that of lexically expressible kind concepts.

9.1 Challenges for the Acquisition of the Conceptual Form of Kind Concepts

The exact nature of the challenges involved in acquiring the conceptual form of kind concepts depends on the assumptions that one makes regarding the innate structure and learning mechanisms available. I begin with considering characteristics of conceptual form that are likely to pose a challenge to any approach.

9.1.1 The Abstractness of Conceptual Form

The perspectives provided by conceptual form cannot be perceived (Prasada, 2021b). We cannot see/hear/smell/feel/taste that Fido is one of an unlimited number of dogs, some of which exist actually and others that exist potentially; that Fido is supposed to have four legs, and if he doesn't, he is not merely atypical but that there is something wrong with him; that Fido has four legs because he is the kind of thing that he is (a dog); and that he counts as exactly one dog, as does every other instance of the kind "dog," whether it is a chihuahua, Great Dane, or any other kind of dog. Nor can the kind that Fido belongs to be sensed; or the fact that certain things are supposed to *not* happen, such as that female lions are not supposed to have manes.

These and other facts about the conceptual form of kind concepts make it clear that the conceptual form of concepts cannot be directly acquired via our senses. Of course, the fact that the conceptual form of kind concepts is abstract and cannot be directly perceived is not a barrier to its being acquired. Even associative learning mechanisms, properly configured, can learn abstract generalizations. For example, they can learn about the types of properties that are relevant to the abstract ontological distinction between objects and substances and use that knowledge to guide the learning of novel names for objects and substances (Colunga & Smith, 2005). Nevertheless, acquisition of the abstract knowledge that constitutes conceptual form may be particularly challenging for the following reasons.

9.1.2 Conceptual Form Is Abstract Along Many Dimensions

As illustrated up to now in this chapter and throughout this book, the conceptual form of even simple concepts such as DOG involves abstraction along multiple dimensions. For example, it involves distinguishing actually existing from potentially existing instances; establishing that instances need only be distinguished numerically, but subkinds and kinds cannot be distinguished only numerically and must also be distinguished by their characters; representing some properties as being true of a kind by virtue of its being the kind of thing that it is; representing those properties as ones that instances of the kind are supposed to have; determining which types of properties have principled connections to the kind for different classes of kinds; representing each instance as one of an unlimited number of instances that are understood to be members of an abstract kind; determining whether the kind individuates instances (e.g., DOG, TABLE), subkinds (e.g., DOG, SAND), or both (e.g., DOG); and determining whether there is a sense in which numerically distinct instances may be understood to be the same and not merely the same kind of thing (e.g., BOOK, NECKLACE), and, in those cases, whether the numerically distinct instances must be perceptually similar (e.g., NECKLACE) or not (e.g., BOOK). As this partial list makes evident, acquiring the conceptual form of kind concepts does not simply involve finding a single abstract generalization that captures the diversity in concrete examples, but rather requires multiple types of abstract formal distinctions and representing multiple abstract formal relations.

9.1.3 The Conceptual Form of Kind Concepts Are Overlapping and Cross-Cutting

A further complication for acquiring conceptual form arises from the variety and intricacy of conceptual forms. That variety and intricacy of conceptual forms poses a particularly difficult acquisition challenge because kind concepts from different classes provide partially overlapping, and sometimes cross-cutting, perspectives. As illustrated earlier (chapter 2), the perspectives provided by LION, FORK, BOOK, and NECKLACE overlap in some respects but diverge in others. For example, all of them allow us to think and talk about an abstract kind as well as an unlimited number of instances of those kinds. They differ, however, in whether they individuate instances in one or more than one way, whether the concrete realizations of abstract instances need to be perceptually similar or not, and whether there are principled differences among instances of a kind that are independent of the subkind identity of those instances (see chapter 2 for a detailed explication of these differences).

The challenges presented by the partially overlapping perspectives that are provided by different classes of kind concepts are exacerbated by the fact that the perspectives provided by kind concepts can often be cross-cutting. For example, MAGAZINE and NEWSPAPER, but not BOOK, provide a perspective in which they allow us to use the term that we use to talk about the type of reading material to also talk about the producer of the reading material (e.g., "The magazine/newspaper/*book hired an editor"). On the other hand, NEWSPAPER, but neither MAGAZINE or BOOK, provides a perspective that allows us to consider all newspapers to be one in a sense, and thus we can say something like "Sally read the newspaper and so did John," even in circumstances in which they clearly read different newspapers (e.g., Sally read *The New York Times* and John read *The New York Post*). On the other hand, a similar statement with "magazine" or "book" (e.g., "Sally read the magazine/book and so did John") can be made only if both people read the *same* magazine or book, even if they read different copies of that magazine/book.

As such, the conceptual form of MAGAZINE is understood to be relevantly similar to NEWSPAPER and different from BOOK in one case, but similar to BOOK and different from NEWSPAPER in the other case. Such cross-cutting perspectives between similar kind concepts are abundant and pose a difficult challenge for acquiring conceptual form.

9.1.4 Conceptual Form Is Not Merely an Abstract Representation of Concrete Properties

There are many models for capturing similarities and differences among concepts within a multidimensional vector space on the basis of the features of different kinds of things (e.g., McRae et al., 1997) or similarity judgments among kinds (e.g., Hebart et al., 2020). Such models provide the means for representing similarities and differences along multiple dimensions and can abstract generalizations about different classes of kinds that arise from the distribution or similarity of the features of instances of kinds from different classes. Such models can identify classes of kinds that can approximate one component of the conceptual form of kind concepts—the component that represents the types of properties to which a given class of kind concepts has a principled connection. Different classes of kinds have principled connections to different numbers and types of properties, and thus kinds within each of those classes are expected to be similar to one another in different ways than kinds in other classes. Such models find the statistical and/or similarity structure present in the properties of different classes of kinds, and thus the structure that is present in the summary representation of those kinds. They do not, however, shed light on the abstract formal structure of the mental lenses that provide instructions for how to think and reason about kinds, instances, and their properties. As already noted, the abstract formal distinctions and formal representations that constitute the conceptual form of kind concepts are not simply abstract descriptions of the concrete properties of instances of kinds. Instead, the conceptual form of kind concepts consists of the mental lenses that specify different ways of thinking and talking about things by providing instructions for how to think about how those properties are related to the kind and its instances.

9.1.5 Minimal and Highly Ambiguous Linguistic Cues to Conceptual Form

Because conceptual form provides instructions for how to think and talk about things and their properties, evidence concerning the conceptual form of kind concepts can come from the ways in which we talk and reason about different classes of kinds, rather than from the properties of those kinds themselves. The linguistic contexts in which names for kinds of things appear can provide important cues about the conceptual form of that kind. That, of course, is the reason why throughout this book, I have been pointing to what it is natural to say (or not say) to motivate, or to provide evidence for, one or another aspect of conceptual form.

Although it is clear that evidence concerning conceptual form can come from the ways in which we talk about different kinds of things, it is not at all clear that distributional information is sufficient for acquiring the conceptual form of kind concepts. Here are some reasons why it is unlikely that we can abstract conceptual form from language alone.

To begin with, there is no unambiguous morphosyntactic way of identifying the conceptual form of any concept or marking any aspect of the conceptual form of a concept. For example, there is no overt way of marking whether a kind individuates instances, subkinds, or both. Similarly, there is no morphosyntactic marker of the fact that instances of a kind can be merely numerically distinct, but kinds cannot be merely numerically distinct and must differ in character; or whether an instance is an actually or potentially existing instance. Even whether or not a noun phrase should be interpreted as referring to a kind cannot be determined solely by the morphosyntactic form of the phrase; it depends on tense, aspect, and world and discourse knowledge (Krifka et al., 1995). The conceptual forms of kind concepts provide intricate and abstract perspectives to think and talk about things and their properties; however, those perspectives are only hinted at by the morphosyntactic structure of language.

Although a combination of morphosyntactic, discourse, and world knowledge cues can help *select* the perspectives provided by any given use of a kind term, those cues do not come anywhere close to *characterizing* the conceptual form. For example, a combination of linguistic and nonlinguistic cues can select a generic (rather than a nongeneric) interpretation of the noun phrase in a sentence like "Dogs bark"; however, those cues do not tell us anything about what such an interpretation amounts to—that it involves characterizing an unlimited number of dogs; that some of those dogs may exist or have existed actually, while others exist only potentially; that instances of the kind are understood to bark because they are dogs; that instances are supposed to bark, and if they don't, something is wrong with them; and that the instances need not be qualitatively distinct (even though they almost always are). All of this, and more, are encoded in the conceptual form of DOG but are not reflected in the linguistic form of the sentence, which allows us to understand the sentence as being about the kind DOG rather than an unspecified group of dog instances. As such, learning the conceptual form of a concept based only on a combination of linguistic cues is unlikely to be successful.

9.1.6 Explicit Characterizations of Conceptual Form Are Likely Rare

One might wonder if more could be learned about the conceptual form of a kind concept by paying attention to the content of what we say about a kind. For example, hearing sentences such as "Barking is one aspect of being a dog," "There is something wrong with a dog that doesn't bark," or "Two dogs can be qualitatively identical and yet numerically distinct" would certainly be helpful for learning the conceptual form of DOG. Such statements are certainly available to readers of this book, but it is almost certain that the statements are completely absent from the natural discourse available to most speakers of English. As such, any viable theory of the acquisition of conceptual form cannot rely on the presence of such statements in the input.

9.1.7 Many Aspects of Conceptual Form Are Acquired Early in Life

Although we need more systematic investigation of when young children acquire various aspects of the conceptual form of kind concepts, it appears that infants may appreciate the link between common nouns and kind representations as early as six to nine months (Campbell & Hall, 2022; LaTourette, Chan, & Waxman, 2023), that ten-month-olds distinguish the perspectives provided by animate kind, object kind, and agent kind concepts (see section 9.4), that two-year-olds distinguish generic and nongeneric utterances (Gelman & Raman, 2003), and that by four years of age, and probably earlier, children appreciate the distinction between properties that have a principled connection to a kind from those that have a statistical connection to a kind (Haward, Wagner, Carey, & Prasada, 2018). These findings suggest that many, if not all, aspects of the conceptual form of kind concepts are available very early in life. If these aspects of conceptual form of kind concepts are acquired via what children hear about the kinds, it will have been acquired on the basis of relatively little input.

9.1.8 Conceptual Form Encodes Nondescriptive Relations

The fact that the conceptual form of kind concepts requires that we represent principled connections between kinds and certain properties poses a further challenge. This means that the acquisition of conceptual form requires representing explanatory links between kinds and certain properties, as well as normative links between kinds and those properties on the basis of descriptive data about what is the case for some number of instances of the kind. Complicating matters further is the fact that though formal explanatory relations are intrinsic to the conceptual form of kind concepts, we also represent causal and teleological explanatory relations between properties of kind concepts. As such, the learner cannot simply assume one type of explanatory relation and seek to determine when to represent such a relation. Instead, the learner must be able to represent each type of explanatory relation and also figure out which ones hold between the kind and different properties. Similarly, principled connections require that we represent those properties that instances of the kind are supposed to have, which can overlap with those properties that we think instances of the kind ideally should have but need not. For example, we can endorse both the statements "Dogs are supposed to have four legs" and "Ideally, all dogs have four legs," but it is odd to say, "#Dogs are supposed to use a litter box," even those it is fine to say, "Ideally, all dogs would use a litter box" (Bublitz & Prasada, 2013). As such, not only must learners be able to represent normative relations between a kind and a property, they must be able to represent the right type of normative relations for a given property and kind. A final complicating factor is that there are many cases in which we represent descriptive, nonexplanatory, and nonnormative connections between kinds and properties, such as TIRES ARE BLACK.

9.2 Two Approaches to the Acquisition of the Conceptual Form of Kind Concepts

Although there are currently no theories of how conceptual form is acquired, it is useful to consider two radically different types of approach that one might take, as well as the problems that they would have to address.

9.2.1 Emergent Perspectives Approach

The *emergent perspectives approach* assumes that there is no innate structure or knowledge concerning conceptual form and the conceptual form of concepts arises via general learning mechanisms that discover the structure present in the instances of kinds and the language used to talk about them. The big challenge for such an approach is whether a learning mechanism that is not in any way specialized for discovering the types of abstractions that are encoded in the conceptual form of kind concepts can discover them. As discussed previously, conceptual form does not simply represent the properties of concrete instances in abstract form; instead, it provides a perspective from which to think and talk about those properties. As such, the acquisition of conceptual form will largely require abstractions and generalizations over ways in which we think and talk about things and their properties.

Recent advances in the capabilities of generative large language models (LLMs) have made it clear that given massive amounts of training data, these models behave in ways suggesting that they have acquired the grammar of the language on which they were trained, and they even behave as if they understand language. An interesting empirical question is the extent to which these models are able to acquire the conceptual form of kind concepts. One reason why they may fail to display knowledge of conceptual form is that the abstract relations encoded in the conceptual form of kind concepts are reflected in the linguistic form and content in a wide variety of ways that have no obvious surface similarities. For example, the principled connection between dog and barking licenses generic uses of "Dogs bark," "A dog barks," and "The dog barks," as well as "Barking is one aspect of being a dog," "Dogs bark because they are dogs," "Dogs are supposed to bark," and "There is something wrong with a dog that cannot bark." Whereas the first set of sentences are likely present in the input, the second set of sentences are likely to be rarely produced in nonexperimental contexts, and the likelihood of the full range of the latter sentences being produced for a given kind and property is even more unlikely. Of course, the input is also likely to contain generic uses of "Dogs wear collars," because of the statistical connection between the kind and the property, but fail to contain generic uses of "A dog wears a collar" or "The dog wears a collar," but also occurrences of "Wearing a collar is one aspect of being a dog," "Dogs wear collars because they are dogs," "Dogs are supposed to wear collars," and "There is something wrong with a dog that is not wearing a collar." As such, it is hard to see how the system may learn to abstract the distinction between principled and statistical connections. Generative LLMs are quite successful at next-word prediction, given the massive amounts of training that they receive, because there are indirect statistical relations between the order of words within and between phrases. This makes next-word prediction and the abstractions that improve it a reasonable learning strategy. In the case of the distinction between principled and statistical connections and other abstract distinctions involving conceptual form, there does not seem to be any obvious relation to next-word prediction. Furthermore, the knowledge encoded in conceptual form is generally presupposed and not articulated. This means that even if there are some very weak next-word patterns that would be evident if the knowledge encoded in conceptual form were articulated, it is unlikely to be present in the training data of even the largest LLMs. Of course, it remains an empirical question whether LLMs display knowledge of the distinction between principled and statistical connections, as well as other characteristics of conceptual form. If they do display knowledge of conceptual form, it will then be important to investigate how much input is required for them to do so and how that compares to the input available to young children.

Recent work by Hu, van Paridon, and Lupyan (2024) provides some relevant initial evidence about the ability of LLMs to learn the distinction between principled and statistical connections. They tested whether eight LLMs that included various versions of BERT (whose name stands for "Bidirectional Encoder Representations from Transformers") and generative pretrained transformer (GPT) could capture the distinction between principled and statistical connections. They found that only GPT-4 succeeded in representing the distinction in a manner that was independent of prevalence. GPT-4 gave higher ratings to sentences that expressed a formal explanation of a property, such as "Because they are airplanes, airplanes have X" when the property had a principled connection to the kind (e.g., *X=wings*) than when the property had a statistical connection to the kind (e.g., *X = seatbelts*). That is impressive—as is the failure of all the other models, including GPT-3.5.

Clearly, with massive amounts of training, it was possible for GPT-4 to learn *one* way in which the abstract formal distinction between principled and statistical connection is expressed in language (licensing formal explanations of the property). It is unknown whether the model also displays knowledge of the many other ways in which the distinction can surface in what we say. Would the model succeed if the formal explanation was expressed in another manner (e.g., "Airplanes have wings because they are the kinds of things that they are" or "Airplanes have wings because they are airplanes")? What if the explanation involved an instance of the kind rather than the kind (e.g., "That has wings because it is an airplane"); or more important, judgments about the normative dimension of principled connections and the many ways in which that normative expectation may get expressed; or linguistic constraints that depend on the distinction (e.g., "The airplane has wings" can be given both a generic and nongeneric interpretation, but "The airplane has seatbelts" can only be given a non-generic interpretation)? For the reasons discussed here, it is unlikely that GPT-4 would automatically display these other abstract characteristics of the formal distinction between principled and statistical connections. The surface forms have no obvious connection to one another, and next-word prediction doesn't seem to be directly related to these various dimensions of principled connections and how they are distinguished from statistical connections. It will be interesting to find out if these models, or more powerful LLMs, can acquire the distinction between principled and statistical connections. Of course, even if they are successful, it would tell us little about how children acquire the distinction, as they acquire it on the basis of only a tiny fraction of the input available to these models.

9.2.2 Perspective Identification Approach

An alternative approach assumes that an intrinsic part of the human conceptual system is a generative system with a limited set of parameters, which yields different classes of lexically expressible kind concepts (see chapter 5 for a sketch of the basic mechanism for representing kinds and some of the parameters in terms of which classes of kind concepts and kind concepts within a class are distinguished). According to this view, we do not have to learn the characteristics of conceptual form because those are intrinsic to the mechanisms in terms of which we form and represent lexically expressible kind concepts. For example, we don't have to learn that kinds are characterized and individuated in terms of the properties that have a principled connection to the kind, that principled connections have an explanatory, normative and statistical dimension, or that instance-of kind representations can differ in their mode of existence. These are formal characteristics of the mechanisms in terms of which we form kind concepts. Furthermore, the learner does not need to somehow discover the various types of perspectives that kind concepts can provide because formally distinct classes of kind concepts and the distinct perspectives that they provide arise from different parameter settings of the kind representation system. For example, the system generates different classes of kinds based on whether the kind individuates instances, subkinds, both instances and subkinds, or abstract and concrete instances. As such, the learner does not have to consider all kinds of other abstract dimensions, such as valence and speed, that can characterize concepts (Liu & Lupyan, 2023) but are not relevant to the perspectives encoded in conceptual form.

Although this approach avoids the question of how we acquire the characteristics of conceptual form and how we can generate the conceptual forms that encode distinct perspectives for thinking and talking about distinct classes of kind concepts, it faces a different acquisition challenge. Because children, and learners more generally, know the abstract formal structures that kind concepts can have, they must figure out which of the innately specified perspectives should be used in forming a concept on a given occasion. So, for example, when children encounter a particular entity, should they conceive of it in a way in which the entity's oneness is understood to be due to its being the kind of thing that it is? Or should they conceive of the entity's oneness as not being due to its being the kind of thing that it is? And should it be conceived of as being a kind of thing where the kind individuates subkinds of that kind? As should be clear, such abstract formal descriptions do not identify the concrete situations to which they should be applied. The learner is faced with what we may call the *perspective identification problem*. How do children know how to apply their innately given formal knowledge of conceptual form and the perspectives that they bring to concrete situations?

That problem is analogous to the problem of how children can identify the innate and abstract categories and structures provided by universal grammar in the specific utterances that they are hearing. Semantic bootstrapping theories propose that innate, unidirectional links between semantic categories such as object, action, and property and syntactic categories such as noun, verb, and adjective allow children to identify the syntactic category of the words used to talk about objects, actions, properties and other elements (Grimshaw, 1981; Pinker,1984). Once the syntactic categories of some words have been identified, children can exploit the abstract knowledge provided by universal grammar to learn parameter settings and surface distributional cues, which will help them identify the syntactic categories of other words and syntactic properties of the language that they are hearing more generally (Grimshaw, 1981; Pinker, 1984). An analogous solution is possible for addressing the perspective identification problem.

9.3 Conceptual Bootstrapping as a Solution to the Perspective Identification Problem

Research over the past three to four decades suggests that our cognitive systems contain core knowledge systems that use abstract principles for identifying and reasoning about entities from a limited number of domains such as objects, places, number, agents, and social beings (Carey & Spelke, 1994; Carey, 2009; Spelke & Kinzler, 2007; Spelke, 2022). Each core knowledge system uses abstract, persisting, and interconnected properties and principles for identifying and reasoning about the entities in their domain. When an infant attends to an entity in the domain of a given core knowledge system, the interconnected properties and principles are activated and guide the infant's expectations about the ways in which the entity may and may not behave. For example, the object core knowledge system uses the principles of cohesion, continuity, and contact for reasoning about the ways in which objects are expected to move: They move cohesively on continuous paths and cannot move in the absence of contact with another object. The agent core knowledge system, on the other hand, represents and reasons about agents in terms of goal-directedness, efficiency, contingency, reciprocity, and gaze direction. Infants expect agents to move in efficient and goal-directed ways toward goals to which they have perceptual access and to interact with other agents in contingent and reciprocal ways. The core knowledge systems are innately given, often are shared with other species, are available throughout our lifespans, and operate unconsciously when we attend to entities in their domain (Spelke & Kinzler, 2007; Spelke, 2022).

Spelke and colleagues understand core knowledge systems to be distinct from both perceptual systems and systems of thought generally. According to their view, core knowledge systems receive their input from perception, and the outputs of core knowledge systems are available to language and combinatorial thought (Spelke, 2022). Scholl (2024) and others have argued that at least some of the knowledge that is proposed to reside in core knowledge systems may instead be found in the representations and mechanisms of midlevel vision and object-based attention or visual working memory (e.g., Leslie, Xu, Tremoulet, & Scholl, 1998; Scholl & Leslie, 1999; Carey & Xu, 2001; Feigenson & Halberda, 2008; Kibbe & Leslie, 2019).

For current purposes, it does not matter which of these ways of thinking about core knowledge systems is correct. What is important for our purposes is the fact that the mind has innately specified systems that are able to represent and reason about different sorts of perceptually given entities in terms of abstract notions that are relevant to their domains, and that those abstract notions are relevant to conceptual forms of different classes of kind concepts.

For example, the agent core knowledge system identifies entities that are capable of self-initiated, goal-directed movement. In so doing, it provides the learner a reason to form a kind concept with the conceptual form of something like DOG. The conceptual form of animate kind concepts such as DOG has a principled connection for the capacity for goal-directed movement; thus forming an animate kind concept such as DOG, RACOON, or LION would explain the property of goal-directed movement that is identified and used by the agent core knowledge system. Although other conceptual forms could also be used to think about the entity (e.g., the conceptual form shared by object kind concepts generally or via the conceptual form shared by living kind concepts), those concepts would leave the property of self-initiated, goal-directed movement unexplained. As such, situations in which the agent core knowledge system is activated privilege some classes of conceptual forms over others for creating a concept for thinking about the thing in question, and thus provide an opportunity to learn some of the perceptual properties and language associated with the kind concepts that explains the properties identified by the agent core knowledge system. Those perceptual cues (e.g., type of form, texture) and language (e.g., lexical items, morphosyntactic properties) may then help the learner identify other situations in which other lexically expressible kind concepts from that class would be relevant for thinking about something, even in the absence of input from the core knowledge system.

The object core knowledge system would privilege the conceptual form of solid substance kind concepts and object kind concepts over other classes of kind concepts. Choosing between the conceptual forms of solid substance kind concepts and those of object kind concepts would require using cues concerning the arbitrariness of the entity's structure. Cues suggesting the structure is nonarbitrary would privilege forming an object kind concept, as the conceptual form of object kind concepts has a principled connection between the object kind and the structure of its instances. The social agent core knowledge system can help identify contexts in which the concept PERSON should be formed.

Exactly which core knowledge systems are available and the exact properties that they identify,[1] as well as the exact classes of kind concepts generated by the generative system for kind representations, will determine how many and which kind concepts may be established in this manner. As children learn more kind concepts in this manner, they acquire more perceptual and linguistic cues to different classes of kinds, which will allow them to increasingly use the language that they have learned and the perceptual cues that are linked to different abstract classes of conceptual forms to learn new kind concepts without direct input from core knowledge systems. My goal here has been to show how core knowledge systems, however they are understood, provide a way for a learner who comes to the task of forming lexically expressible kind concepts with knowledge of the abstract characteristics of conceptual form, and the classes of possible conceptual forms for kind concepts can rationally determine when a kind concept from some of those innately given classes should be established. Furthermore, the core knowledge systems help establish some of the properties that have principled connections to the kind, and thus establish an important part of the kind syndrome by which new instances of the kind may be identified (Block, 1986; Macnamara, 1986; Margolis, 1998).

It is worth noting that because the conceptual forms of kind concepts from specific classes *explain* the properties identified by specific core knowledge systems, the link between core knowledge systems and lexically expressible kind concepts is neither arbitrary nor stipulative. As such, the conceptual bootstrapping solution for the perspective identification problem avoids McPherson's (1995) criticism of semantic bootstrapping theories for stipulating links between specific semantic categories and specific syntactic categories.

9.4 Perspectives Available to Infants

Adult conceptual systems possess kind concepts that display a wide variety of abstract perspectives. Which of those perspectives characterize infants' earliest concepts? A recent review of the literature by Croteau, Cheries, and Xu (2024) suggests that by ten months, infants distinguish the perspectives provided by object kind concepts, animate kind concepts, and agent kind concepts. The evidence that infants distinguish the perspectives provided by kind concepts from these three classes of kind concepts comes from studies that show that infants can individuate an instance of a kind from one of these classes of concepts from an instance of a kind from a different class, but not instances of two kinds within each of the classes. Evidence for that comes from experiments using "Is-it-one-or-two" tasks. In such tasks, infants see a screen on a stage and then see one item (e.g., a ball) appear from behind the screen before returning behind the screen. They then see a second item (e.g., a duck) emerge from behind the screen before returning behind the screen. Finally, the screen is removed to reveal either one of the objects or both objects, and looking time is measured for each of those outcomes. If infants form distinct instance-of-kind representations for the two items, they should expect there to be two items behind the screen and thus look longer when the screen is lifted to reveal only one item. If children are unable to form distinct instance-of-kind representations for the two items, the outcome in which two items are revealed should be unexpected and draw more looking.

Reviewing a large body of studies, Croteau et al. (2024) find that infants younger than ten months successfully individuate the items when the two items are instances of kinds from distinct superordinate categories such as animate kind, object kind, or agent kind, but not when both items come from the same superordinate category. Croteau et al. interpret those results as suggesting that infants have the superordinate sortal-kind concepts ANIMATE, OBJECT, and AGENT, but not basic-level concepts like DUCK, CAT, CUP, and BALL. The theory of conceptual form suggests a different interpretation.

Because the theory of conceptual form posits a distinction between a concept's form and content, differences in how we think and talk about things from different domains can be understood as arising from differences in the conceptual form of basic-level concepts without having to represent superordinate concepts. For example, we think of ducks and balls in different ways because the conceptual forms of DUCK and BALL are different and instruct us to think of their content in different ways. As such, the results that Croteau et al. (2024) review are consistent with the idea that by ten months of age, infants distinguish the perspectives provided by animate kind concepts such as DUCK, object kind concepts such as BALL, and agent kind concepts such as PERSON, but they have not yet figured out which property differences distinguish kind concepts that have the same conceptual form. That can be a difficult thing to know in many circumstances, even for an adult. For example, what property differences distinguish instances of DOG from instances of DINGO? On the other hand, the capacity for self-initiated motion, or lack thereof, easily supports the formation of a different instance-of-kind representations when encountering a duck and a ball because DUCK and BALL have different conceptual forms, and thus their instance-of-kind representations are formally distinguished in addition to being distinguished by their content.[2]

Given this interpretation of the results reviewed by Croteau et al. (2024), it appears that by the time children begin learning their first words, they have access to at least three classes of kind concepts, which provide different perspectives for individuating and thinking about their instances.

9.5 How the Form and Content of Core Knowledge Systems Differs from That of Lexically Expressible Kind Concepts

Spelke (2022) suggests that language may be a system by which the contents of different core systems may be combined. That proposal may tempt some to wonder whether the representations made available by core knowledge systems may be the representations onto which words are mapped. For example, can the meaning of "duck" be represented via the representation that the agent core knowledge system generates when we watch a duck engaged in self-initiated motion? There are several reasons to resist this possibility, even though language may provide a means for combining the contents of different core knowledge systems.

First, core knowledge systems, like perceptual systems, generate representations of *particulars*. As such, the representations do not represent the entities as one of an unlimited number of instances of a given kind. They also do not represent the kind to which the particular entities that they represent are understood as belonging, and, more generally, the kind and instance-of-kind perspectives provided by lexically expressible concepts like DUCK. Even though core knowledge system representations have content that is relevant to the content of lexically expressible kind concepts, the content is limited in ways that the content of lexically expressible concepts is not. Second, Carey (2009) suggests that the contents of core knowledge systems are represented in an iconic format (but see Green & Quilty-Dunn, 2021). That would make the format of core knowledge representations ill suited to being directly combinable via language.

Core knowledge systems identify characteristics that the conceptual form of kind representations (of the relevant class) represent as being aspects of being that kind of thing, and thus to have a principled connection to the kind. In doing so, the conceptual form of the kind concept not only provides a format that allows the content of the core knowledge representation to be linguistically expressed and combined with other concepts via language, it also transforms the content of the core knowledge representation into the type of content that is expressed via lexically expressible kind concepts, which provide perspectives of the sort discussed throughout this book. The representations within core knowledge systems do not provide those perspectives. It thus appears to be the case that core knowledge representations can help identify/construct lexically expressible concepts that share some of the content of core knowledge representations. Our linguistic systems can use those lexically expressible concepts, but not core knowledge representations, to combine the contents of core knowledge systems and much else.

V Conclusions

10 The Nature of Conceptual Knowledge

As we have seen, lexically expressible concepts provide intricate and abstract perspectives for thinking and talking about things and their properties. The perspectives provided by a concept are not properties of the things that the concept is used to think and talk about. Instead, they are the "lenses," or sets of instructions for thinking and talking about things when using a given lexically expressible concept. Those perspectives are encoded in the formal structure of the concept, its *conceptual form*, which provides instructions for how to think and talk about things and their properties when using that concept and helps explain a range of linguistic and nonlinguistic phenomena. Furthermore, lexically expressible concepts can be used to build arbitrarily complex concepts by combining them and expressing them via language. Our faculty for commonsense conception is partly constituted by a generative system that yields the conceptual forms of a wide array of lexically expressible concepts that vary in terms of a limited number of formal components and the manner in which they are realized. A fragment of that system has been presented in the preceding chapters, especially chapters 4 and 5. An important task for future research is to expand the breadth of the concepts that can be generated, as well as the depth of some of the concepts generated by the fragment of the system so far characterized.

Here, I step back to reflect on what the theory of conceptual form reveals about the nature of conceptual knowledge by considering key ways in which it differs from existing approaches to conceptual representation. Because the theory of conceptual form provides a radically different way of thinking about concepts, it potentially provides new ways of thinking about many foundational questions about conceptual representation and also identifies some questions and issues that have not been discussed or were not formulable in standard approaches. Nine key characteristics of the theory of conceptual form that distinguish it from most existing work on concepts are the following:

1. Kind concepts represent both shared characteristics *and* how instances differ.
2. Instance-of-kind representations are intrinsic to kind representations (for most kinds).
3. Certain types of general knowledge can be acquired via parameter setting.
4. Principles are *acquired.*
5. Conceptual form embodies sortal principles.
6. Lexically expressible kind concepts use kind-determined modes of explanation and description.
7. Lexically expressible kind concepts are conceptual primitives.
8. Lexically expressible kind concepts are primitive, but not innate.
9. Domain specificity arises from parametric variation allowed by domain-general principles.

Although these nine ways in which the theory of conceptual form differs from other approaches to conceptual representation are not independent of one another, they help elucidate the nature of conceptual knowledge encoded in the conceptual form of commonsense kind concepts and the ways in which that differs from other approaches to conceptual representation. It is important to keep in mind that in many cases, the theory of conceptual form and other approaches capture *complementary* aspects of conceptual knowledge. As such, difference need not imply incompatibility. My goal in this chapter is to provide prima facie reasons for why the theory of conceptual form suggests the differences identified here and discussed next, rather than present detailed arguments for each difference, as that would require a chapter or more for each of the points. Perhaps this chapter will serve as an impetus for developing such arguments and counterarguments in future research.

10.1 Kind Concepts Represent Both Shared Characteristics *and* How Instances Differ

Most existing approaches to conceptual representation represent lexically expressible kind concepts such as DOG, TABLE, WOOD, BOOK, and LION via summary descriptions, which represent the characteristics that are *shared* by instances of the kind. Approaches differ about whether the summary description is a definition that represents the individually necessary and jointly sufficient properties shared by any and all members of the kind; the summary description is a prototype that represents the properties that are typically shared by instances of the kind; or the summary representation consists of the causal-explanatory structure that is shared by instances of the kind.

But, as we have seen, kind representations also need to contain information that specifies how instances of a kind are understood to *differ*. For example, we understand that the distinctness of two instances of WOOD is not due to their being wood, but the distinctness of two instances of TABLE or DOG is due to their being the kinds of things that they are; that two instances of DOG may be merely numerically distinct and thus have no qualitative differences; that we distinguish and count instances of BOOK in two ways; and that some properties, such as the presence/absence of manes, display principled variation within instances of a kind like LION. Furthermore, as discussed in chapter 7, patterns of systematic polysemy are licensed by the ways in which different classes of kinds *individuate* their instances and subkinds. These, and other similar facts, show that a key component of a kind representations is to specify instructions for how to think about how instances of the kind *differ* and how the distinctness of an instance is related to their being the kind of thing that they are.

Standard approaches to conceptual representation do not have the means for capturing such facts because they represent kinds via summary descriptions that contain only the *shared* characteristics of the kind, and thus the ways in which instances of the kind are thought to be the same (in the kind of thing that they are) or similar (in their characteristics).

The conceptual form of kind representations, on the other hand, provides instructions for how to think about shared properties, *as well as* how instances and subkinds of that kind differ. Furthermore, different classes of kinds are distinguished by the principles of individuation that they provide. Kind representations are generative mechanisms with a component that represents the character of the kind via the properties that have principled connections to the kind and are understood to be aspects of being that kind of thing. That component also optionally represents properties that characterize the kind via causal and statistical connections to the kind. That component is analogous to the summary descriptions found in standard approaches, in that it represents the properties that generally characterize instances of the kind. Importantly, that is only one component of kind representations. Kind representations also generate instance-of-kind and subkind representations and specify formal characteristics of how each of them is distinguished from one another and is related to the kind. Standard approaches have no counterpart for this component of kind representations. The conceptual form of kind concepts provide perspectives for thinking and talking about the principled ways in which instances and subkinds of a kind are the same, as well as the principled kind-determined ways in which they differ.

10.2 Instance-of-Kind Representations Are Intrinsic to Kind Representations (for Most Kinds)

We think and talk about particular things such as the table in the corner, my neighbor's dog, or the sand in my shoe. In doing so, we identify those particular things as instances of one or another kind. Despite the massive literature on categorization, surprisingly little has been explicitly written about the mental representations that represent the particulars that we encounter as an instance of a kind/token of a type (some notable exceptions are Jackendoff, 1983, 1987, 2002; Barsalou 2012). The little literature that there is tends to assume that things are identified as a token of a type by "activating" or combining two mental representations, a TOKEN representation and a TYPE representation, via predication or some other means. The TOKEN representation is understood to be an untyped representation of a particular—it represents a particular thing without specifying its type, and the TYPE representation is understood to be a description of the shared features of the instances, lacking information about what distinguishes instances. That approach, which we may dub the *Identification via Combination* approach, differs radically from the approach provided by the theory of conceptual form, in which a single instance-of-kind representation that is generated by the kind representation provides the means for identifying particular things as instances of kinds.

The type/kind representations that are assumed by the Identification via Combination approach contain only descriptions of the shared properties of instances of a kind and thus are problematic because kind representations must also specify the principled kind-determined ways in which instances and subkinds of a kind are understood to differ. Here, I want to consider a different problem that the Identification via Combination approach gives rise to.

Any account that captures the identification of something as an instance of a kind by combining two distinct representations via predication, coactivation, or some other way of applying one mental representation to the other fails to represent the intrinsic unity that we understand an instance of a kind to possess. My neighbor's dog is not understood by us to be two things at once—a thing of an indeterminate or unidentified sort that also happens to fit the description of a dog at this time. Instead, we understand it to be a single thing—a dog. Furthermore, we expect it to remain a dog—in fact, to remain the same dog throughout its existence. Identification via Combination approaches must specify for how long, and in which contexts, the things represented by each representation should be combined. Furthermore, if the mode of combination allows for combining things that are accidentally united as well as things that are essentially or lawfully united, then the appropriate mode of combination must also be specified.

The theory of conceptual form, on the other hand, avoids all such problems because a *single* instance-of-kind representation provides the means for identifying something as an instance of a kind. The single instance-of-kind representation, which is generated by kind representations like DOG and TABLE, has a form that specifies both kind and instance identity, and thus provides a perspective for thinking and talking about things as instances of kinds. The instance-of-kind representations generated by kinds like DOG and TABLE are primitives that do not involve combining concepts. As such, no questions arise about the length of time or circumstances in which an untyped TOKEN is understood to be connected to a given TYPE. The unity of instance and kind identity is intrinsic and primitive and thus does not require combining representations.

There is, however, a class of kinds that does not individuate instances. For example, SAND does not individuate instances, so it must be combined with an individuating concept such as PILE in order to identify an ad hoc instance of SAND. Instance representations generated in this manner cannot be distinguished from other instances of the kind in a fixed manner and can even be transformed into other instances of the same kind (Prasada & Hall, 2019). Instances of intrinsically individuated kinds like DOG and TABLE, on the other hand, are understood to remain the same instance throughout their existence and cannot be transformed into another instance of the same kind (Prasada & Hall, 2019).

The Identification via Combination approach, which appears to be implicitly or explicitly adopted by existing approaches to conceptual representation, has a uniform way of representing instances of kinds/tokens of types. On that approach, representing something as an instance of a kind requires combining token and type representations for all types of things. As such, the way that it represents things as instances of kinds fails to capture the different ways in which we think about the identity of instances from kinds that individuate instances (e.g., TABLE) and kinds that do not (e.g., SAND). An advantage of the theory of conceptual form is that it captures those differences in the form and generation of the instance representations in the two types of cases.

10.3 General Knowledge Can Be Acquired via Parameter-Setting

Much of our world knowledge consists of knowledge about kinds of things despite the fact that we never encounter kinds of things. We only encounter particular things, or sets of particular things that are instances of various kinds of things. How, then, do we acquire knowledge about kinds of things? A standard answer is that knowledge about kinds of things is acquired via a process of induction whereby we first acquire knowledge about particular things and then, based on our knowledge of those particular things and other innate or acquired biases or constraints, we make a generalization about what instances of that kind are generally like. This tenet of conceptual empiricism—that knowledge of particulars precedes knowledge of the general and knowledge of the general is acquired via induction—is widely shared by researchers that nevertheless differ with respect to the existence and specificity of innate constraints on induction.[1] A second tenet of conceptual empiricism is that knowledge is initially descriptive, and only later explanatory. Keil, Smith, Simons, and Levin (1998) note that this tenet is manifested in theories of concept formation that rely on theory-neutral computations of similarity initially, and only later structure concepts in terms of explanatory principles. A third tenet of conceptual empiricism is that we begin with descriptive knowledge, and thus any normative knowledge that we come to have poses a difficult problem—how can we go from what we know *is* the case to conclusions about what is *supposed to be* the case, or what *ought to be* the case? This tenet motivates research into the biases or factors that underlie a descriptive to prescriptive shift (e.g., Tworek & Cimpian, 2016; Roberts, 2022; Foster-Hanson & Lombrozo, 2022).

The conceptual form of kind mechanisms contradicts each of these tenets of conceptual empiricism. To see why, let us briefly recap the key components of the conceptual form of kind representations. Recall that because kinds are understood to contain an unlimited number of instances, they are represented via a generative mechanism that is capable of generating an unlimited number of instance-of-kind representations. Furthermore, the conceptual form of kind representations contains a component that characterizes the kind and distinguishes it from other kinds via the properties that have principled connections to the kind. These properties are understood to be aspects of being that kind of thing, and thus are properties that instances of that kind are supposed to have by virtue of their being the kinds of things that they are. Consequently, the presence of these properties in instances of the kind are explained by the instances being the kinds of things that they are.

Examining the relation between particular and general knowledge in the conceptual form of kind representations, we see that the instance-of-kind representations, which provide the means for storing knowledge about particular members of a kind, are formally *dependent* on the kind representations that generate them. As such, the representation of the particular is not prior to the representation of the general. Kind representations and instance-of-kind representations are mutually defined and generally come in together, but as noted in chapter 3, we can have kind representations without any instance-of-kind representations. For example, we can design a new kind of artifact, the watermelon pulverizer, and think and talk about this new kind of artifact even when no instances of the kind exist.[2] Furthermore, the properties that have a principled connection to the kind are the properties that instances are supposed to have by virtue of their being instances of that kind. As such, our expectations about what properties particular instances of a kind are supposed to have is determined by the kind of thing that it is. Once again, our knowledge of the particular (instance) is not prior to our knowledge of the general (kind). Instead, we have expectations about what properties an instance of a kind will have on the basis of our knowledge of the kind of thing that it is, in advance of any other knowledge about the particular instance.

Finally, we characterize what a kind is like in terms of properties that its instances are *supposed to* have and that are *explained* by its being the kind of thing that it is. The descriptive, normative, and explanatory are intrinsically intertwined in principled connections. The conceptual form of kind representations characterizes kinds and their instances in normative and explanatory terms in the course of describing what they are like.

All of this makes clear that the conceptual form of kind representations contradicts each of the tenets of conceptual empiricism identified previously. Furthermore, recent research by Haward, Carey, and Prasada (2021) provides empirical evidence that the conceptual form of kind representations can guide the formation of kind representations and reasoning about instances of kinds, starting with the first instance of a kind that we encounter. In their experiment, Haward et al. presented participants with a picture of a novel animal or artifact and taught them a novel name for the kind (e.g., "This is a *blick*") and a fact about the instance using a locution that suggested that the property had a principled connection to the kind (e.g., "It is purple because it is a *blick*") or a locution that suggested that it is a common property of *blicks*, but which did not give any evidence that the property had a principled connection to the kind (e.g., "Almost all *blicks* are purple"). Participants were then shown another instance of the novel kind and told that it lacked the property that they had been taught about using the first instance (e.g., "Here's another *blick*, and it is not purple"). Finally, participants were asked about the normative implications of the second instance lacking the property (e.g., "Does that mean there is something wrong with this *blick*?"). If participants had learned the property using a locution that suggested the property had a principled connection to the kind, they were significantly more likely to say that there was something wrong with the instance that lacked the property than if they were not given any information that the property had a principled connection to the kind. Importantly, participants' response that there is something wrong with the *blick* missing the property cannot be due to their simply viewing the item as being atypical, as participants were significantly less likely to say that there is something wrong with a *blick* missing the property after being told that almost all *blicks* have the property than after being given evidence that it had a principled connection to the kind.

These results show that if participants are provided with evidence that a property has a principled connection to a kind via a descriptive fact (e.g., "being purple is part of being a *blick*") or an explanatory fact ("It is purple because it is a *blick*"), participants assume a normative connection between the kind and the property, even though they were given no direct evidence for the normative connection. This provides evidence that the descriptive, explanatory, and normative are intertwined in principled connections and that, contrary to the assumptions of conceptual empiricism, the descriptive does not have priority over the normative or the explanatory in these situations.

In a follow-up study, using a slightly simplified method, Haward (2020) found a similar result with four year-olds. The results with the children provide strong evidence that the descriptive, explanatory, and normative are intrinsically connected in principled connections and are available for structuring our kind representations from an early age. The results of both studies also show that we can set up expectations about whether instances of a given kind are supposed to have certain properties on the basis of experience with a single instance, given the right linguistic evidence. This means that it is not necessary to learn about a number of particulars (instances) before we can acquire general knowledge (knowledge about the kind).

Together, these results show that there is an important class of generic knowledge that is not acquired via a process of induction, but instead is acquired via a process of selection and parameter setting (see chapter 6) that can be linguistically guided. The knowledge acquired in this manner is intrinsically general, normative, and explanatory. As such, conceptual empiricism cannot provide a theory of conceptual knowledge in general.

It is important to note that I am not arguing that we *never* acquire general knowledge in the manner described by conceptual empiricism, only that the theory of conceptual form shows that there is an important class of generic knowledge that is acquired in a manner that is at odds with the tenets of conceptual empiricism. In fact, it is likely that generic knowledge that involves statistical connections such as the knowledge that tires are black is likely to be acquired in the manner described by conceptual empiricism. However, as discussed in chapter 6, generic knowledge involving statistical connections presupposes at least some generic knowledge that involves principled connections and thus cannot be acquired prior to generic knowledge involving principled connections.

10.4 Principles Are *Acquired*

Having considered how the theory of conceptual form challenges standard assumptions about the manner in which general knowledge is acquired, I want to turn to considering the nature of the knowledge that is acquired via the mechanisms for kind representation, in particular generic knowledge involving principled connections such as the knowledge that dogs bark, cars have wheels, and Mormons don't drink alcohol. In acquiring knowledge of this sort, we acquire *principles* for how to think about kinds and their instances. The principles provide starting points for thinking about kinds and their instances. They specify what one has reason to think about the kind and its instances simply because the object of thought is that kind of thing. The principles that we acquire are intrinsically general—they apply to an unlimited number of instances; normative (they specify what properties instances of a kind are supposed to have) and explanatory (principles specify properties instances have because they are the kinds of things that they are).

Although many researchers acknowledge the role of innate principles in guiding the acquisition of conceptual knowledge (e.g., Keil, 1989; Gelman, 1990; Caramazza & Shelton, 1998; Carey, 2009), it is usually assumed that those principles guide the acquisition of knowledge, which is always open to revision (Carey, 2015). As discussed earlier, the principles about kinds that we acquire are unrevisable. Because the principles have an intrinsic normative dimension that specifies properties that instances are supposed to have by virtue of being the kinds of things that they are, the existence of instances lacking those properties cannot count as evidence against the principle. What is the case cannot count as evidence against what is supposed to be the case. For example, the fact that some dogs do not have four legs does not in any way impugn the validity of the principle that dogs, by virtue of being dogs, are supposed to have four legs. The potential divergence between our acquired principles and what is the case is neatly captured in the coherence of statements like "In principle, all dogs have four legs, but, in fact, it is likely that not all dogs have four legs." What is understood to be true in principle can sit comfortably with what is true in fact, making the principles that we acquire about kinds of things unrevisable.

The formal mode of explanation involved in the representation of knowledge involving principled connections provides a second reason why such knowledge is unrevisable. Formal explanations are licensed by the property being represented as an aspect of being the kind of thing in question. As such, the connection between the property and the kind that explains it is unmediated. It is because the property is represented as an aspect of being a given kind of thing that it can be explained by reference to the kind. In contrast, causal explanation posits the existence of some (often unknown) mechanism that explains why the kind and property are connected. As such, the causal mode of explanation represents the connection between the kind and the property as being mediated rather than immediate, and thus allows for the possibility that we can change our mind about what mediates that connection.

Work on conceptual representation and development has generally focused on the acquisition of factual (including explanatory) knowledge, and thus has correctly assumed that any such knowledge that we acquire is revisable. In contrast, the theory of conceptual form suggests that lexically expressible kind concepts, at their core, provide principles for thinking and talking about kinds and their instances. The acquired principles include sortal principles, as well as the principles that characterize each kind. On this view, concept formation is primarily the acquisition and elaboration of principles, and only secondarily the acquisition of revisable factual knowledge. Principles may be abandoned or adopted, but they cannot be revised. This can help explain why the intuitive understandings embodied in the principles that constitute our lexically expressible concepts (e.g., bubbles are weightless) persist even after we have been taught scientific theories that conflict with those intuitive understandings (Goldberg & Thomson-Schill, 2009; Shtluman & Valcarcel, 2012; Kelemen, Rottman, & Seston, 2013; Shtulman & Young, 2024). If, for example, we form the understanding that bubbles, by virtue of being bubbles, are weightless, we cannot revise that principle on the basis of scientific instruction. We can, however, learn to privilege the understanding provided by the scientific theory. As Shtulman and Young (2024) put it, "Scientific interpretations of a domain can be prioritized over intuitive ones but the conflict between science and intuition cannot be eliminated altogether." This persisting conflict is explained if the early understanding is embodied in principles that have an intrinsic normative dimension and involve formal explanation, and thus are unrevisable, but not if our initial understandings are represented in nonnormative descriptive terms and explanatory notions that are revisable.

10.5 Conceptual Form of Kind Representations Embody Sortal Principles

The need for sortal concepts has been a source of debate in both philosophy and psychology (Ayers, 1997; Gupta, 1980; Macnamara, 1986; Xu, 1997, 2010; Carey & Xu, 1999; Rips, Blok, & Newman, 2006, Leonard & Rips, 2015; Wiggins, 1980, 1997, 2001; Rhemtulla & Xu, 2007; Blok, Newman, & Rips, 2007; Hall & Waxman, 1993). The theory of conceptual form provides a new perspective for thinking about the nature and need for sortal concepts. Specifically, the principles that sortal concepts provide constitute a core part of the perspectives provided by lexically expressible kind concepts. These principles are embodied in the conceptual form of kind concepts and guide the way that we think and talk about kinds and their instances. This internalist interpretation of sortal concepts and principles has many benefits and, among other things, helps highlight the nonreductionist perspectives that lexically expressible noun concepts provide for thinking and talking.

Drawing on the work of philosophers, John Macnamara (1982, 1986, 1994, 1999) argued that acquiring the meanings of count nouns like "dog," "bottle," and "tree" and mass nouns like "milk," "sand," and "plastic" requires children to have access to sortal concepts. Sortal concepts are understood as providing a principle of application, a principle of identity, and sometimes a principle of individuation. The *principle of application* specifies what the concept "applies to," or lets us think and talk about, and is provided by sortal concepts such as DOG, TABLE, and SAND, as well as nonsortal concepts such as SING, TALL, and BETWEEN. Sortal concepts apply to kinds—they allow us to think and talk about kinds of things. So DOG allows us to think and talk about the kind "dog," and SAND allows us to think and talk about the kind "sand." On the other hand, nonsortal concepts like SING, TALL, and BETWEEN do not provide the means for thinking and talking about a kind. They instead provide the means for thinking and talking about an activity, a property, and a relation, respectively.

The *principle of identity* specifies the respect in which an instance of a kind remains the same throughout its existence. Fido, for example, is understood to remain the *same dog* throughout his existence. He can change in a multitude of ways while remaining the same dog. For example, he does not remain the same set of cells, or the same weight or height, and his temperament may change, and yet we understand him to remain the same dog throughout his existence. Similarly, we can sensibly say that sand currently sitting in a cup now is the same sand that was in the bowl earlier today, or that the sand that was wet earlier is now dry. The availability of a principle of identity does *not* require us to be able to correctly decide whether, for example, a given dog is the same dog as the one we saw at another time, or if the sand in the cup now is the same sand as the sand that was in the bowl yesterday. Whether we are able to correctly make that identification or not, the fact that DOG and SAND provide principles of identity allows us to think about the identity of instances of those kinds, and, for example, wonder whether the sand in the cup now is the same sand as was in the bowl yesterday. We cannot do the same for nonsortal concepts. It makes no sense to think or say "#Is that the same sing/tall/between as the sing/tall/between we saw yesterday?" (Baker, 2003). The concepts SING, TALL, and BETWEEN do not specify a way of being the same throughout existence. Note that the principle of identity provided by sortal concepts is nonreductionist.

The principle of identity allows us to think that instances of a kind will remain the same instance of that kind—the same dog, the same sand. The principle does not give a reductionist account of what it means to remain the same dog or the same sand. That is, the principle does not specify some set of nondog properties that amount to being a dog and would have to persist in the proper manner in order for something to remain the same dog. The principle captures the fact that our conceptual systems treat instances of kinds as having a basic, nonderived form of continuity as instances of that kind. For example, I know that Fido will remain the same dog that he has always been, but this does not mean that I think he will be the same set of cells throughout his lifetime, or that his being the same dog merely amounts to the matter that he is constituted of displaying spatiotemporal or causal continuity throughout his lifetime. I may think that in addition to thinking that Fido remains the same dog throughout his lifetime, he will display some other forms of continuity, but I need not think that, nor do I need to think that those forms of continuity are equivalent to being the same dog. Being the same instance of a kind is not understood to be reducible to being the same something else.

Finally, count sortals like DOG, BOTTLE, and TREE provide a *principle of individuation,* but mass sortals like MILK, SAND, and PLASTIC do not. The principle of individuation specifies what counts as one instance of the kind, and thus a unit of quantification for the kind. Kinds like "milk," "sand," and "plastic" do not specify what counts as one instance of those kinds; and instances of those kinds must be identified on an ad hoc basis by combining the kind concept with an individuating concept such as PILE, HEAP, PORTION, or PIECE, or via other linguistic and nonlinguistic means (e.g., "The sand in my shoe"). In contrast, kinds such as "dog," "bottle," and "tree" *do* specify their instances. Although instances may differ in their qualities and sizes, each one counts as exactly one instance of the kind. A chihuahua, a Labrador, and a Great Dane each count as exactly one dog. And each bottle counts as exactly one bottle, no matter its shape, size, material, color, or any other quality. We also naturally think of there being an unlimited number of quantitatively equivalent instances of each count sortal.

Traditional sortal theories simply distinguish between kinds that provide a principle of individuation that individuates their instances (e.g., dog) and kinds that do not provide a principle of individuation for individuating their instances (e.g., sand). This captures an important difference in the perspectives provided by kind concepts of two types; however, as we saw in chapter 5, the perspectives provided by lexically expressible kind concepts are, in fact, more intricate and variegated than suggested by traditional sortal theory. Concepts like DOG not only individuate instances of the kind, but also subkinds of the kind; concepts like SAND do not individuate instances but do individuate subkinds; concepts like TABLE intrinsically individuate instances but not subkinds; and concepts like BOOK individuate instances in two ways.

Although kind concepts in these classes tell us whether their instances and/or subkinds are intrinsic units of quantification or not, they do not provide us with infallible criteria for identifying instances or subkinds. For example, the fact that FORK individuates its instances means that any two things identified as instances of FORK must be counted as quantitatively equivalent, unlike, for example, any two instances of PLASTIC. This does *not* mean that the concept provides some infallible way of identifying instances of FORK. Is a fork with one tine missing an instance of FORK, albeit a damaged one? Or is it just part of a fork? What if two tines are missing? Or three of four tines are missing? Or four of four tines are missing? The fact that FORK provides a principle of individuation does not mean that the concept provides clear answers to those questions (which really pertain to the applicability of FORK to an entity). What it *does* mean is that if, for example, a fork with two tines missing is identified as an instance of FORK rather than as an instance of PART OF A FORK or an instance of PLASTIC, it counts as *one* fork, and it is quantitatively equivalent to an instance of FORK that has all four of its tines (and all other instances of FORK). In contrast, assuming that they are plastic forks, we cannot assume that those two instances of PLASTIC are the same quantity of plastic. Consequently, difficulties in determining what to identify as an instance of a kind for the purposes of counting (Shipley & Shepperson, 1990; Srinivasan, Chestnut, Li, & Barner, 2013; Leonard & Rips, 2015; Syrett & Aravind, 2021) cannot on its own be used as evidence against the concept providing a principle of individuation.

In addition to providing principles of application, identity, and individuation (of various sorts), all lexically expressible kind concepts provide a principle of explanation that explains properties with a principled connection to the kind in terms of the kind of thing that something is, and a normative principle that specifies which properties instances of a given kind are supposed to have. We thus see that the sortal principles that have traditionally been discussed are an important, but limited, part of the rich perspectives that are provided be lexically expressible kind concepts for thinking and talking.

One objection to sortal theories is the fact that we are able to understand fairy tales in which instances of one kind change into instances of another kind (e.g., Rips, Block, & Newman, 2006; Leonard & Rips, 2015), such as a prince being turned into a frog or a pumpkin into a carriage. If concepts like FROG and PUMPKIN provide a principle of identity, the objection goes, such transformations should not be comprehensible and should be thought to be impossible. The comprehensibility of such transformations would indeed be problematic for sortal theories if the principle of identity were represented as a *descriptive universal generalization* such as (84). If, on the other hand, it is instead understood to be a *principle* (85), and represented in the same manner as other principles that guide our thinking about kinds are represented (86), then the existence of such transformations is not problematic:

(84) All pumpkins remain pumpkins across all times and situations.

(85) Pumpkins, by virtue of being pumpkins, remain pumpkins throughout their existence.

(86) Dogs, by virtue of being dogs, are four-legged.

Given the principle in (86), simply knowing that something is a dog, in the absence of any other knowledge about that instance or its circumstances, gives us a reason to expect the dog to have four legs. If it turns out that the dog does not have four legs, we think that is not what is supposed to be the case. Similarly, given the principle of identity in (85), simply knowing that something is a pumpkin, in the absence of any other knowledge about the instance or its circumstances, gives us a reason to think it will remain a pumpkin throughout its existence. Furthermore, if it turns out that the pumpkin does not remain a pumpkin, we think that is not what is supposed to be the case.

Given the principle in (86), the four-leggedness of a dog is explained by its being a dog, but a dog not having four legs is not explained by its being a dog (Rivera, Prasad, & Prasada, 2023), but by something other than its being a dog. Similarly, given the principle of identity in (85), a pumpkin remaining a pumpkin throughout its existence is explained by its being a pumpkin, but the transformation of a pumpkin into a carriage is not explained by its being a pumpkin, but by something other than its being a pumpkin—some kind of magical event.

The principle of identity not only allows violations, but it also explains the form of continuity found in instances that do not violate the principle and guides our thinking about the instances that do violate the principle. Furthermore, as discussed in chapter 4, sortal information constrains our thinking about which transformations are possible (Prasada & Hall, 2019). Although we can entertain the possibility of an actually existing instance of one kind (e.g., raccoon_{27}) being turned into a currently potentially existing instance of a different kind (e.g., beaver_{58}), we do not think that it is possible to transform it into a different potentially existing (raccoon_{96}) or actually existing (raccoon_{21}) instance of the original kind or an actually existing instance of a different kind (e.g., beaver_{23}). This pattern of thinking is explained by the conceptual form of object kind concepts.

The principle of identity, and sortal principles more generally, are an important component of the perspectives provided with lexically expressible kind concepts. The conceptual form of these concepts encodes a perspective in which what something is is primary, in that it provides the starting point for thinking, talking, and reasoning. What something is explains why things have certain properties, specifies what properties it is supposed to have, specifies what we expect to generally be the case concerning those properties, specifies one or more units of quantification, specifies a form of continuity through time and change, and can specify a frame of reference for thinking and talking about orientation and direction in space (i.e., what counts as front/back and upright for an instance of that kind).

This perspective, which is encoded in the conceptual form of kind representations, is one in which the world is not a blooming, buzzing confusion but instead an intelligible world that contains instances of various kinds. We think, talk, and reason about those things and understand their qualities, quantities, places, and change in terms of the kinds of things that those things are. This kind-centric, nonreductionist perspective is at odds with the perspective assumed by standard approaches to conceptual representation. Standard approaches take a reductionist approach, in that they seek to represent lexically expressible kind concepts such as DOG, BOTTLE, and SAND in terms of kind-independent representations of qualities, quantities, space, time, and causal processes.

Our conceptual systems allow both ways of thinking and reasoning; however, it is the nonreductionist, kind-centric, formal mode of understanding encoded in the conceptual form of kind representations that is relevant to the morphosyntactic constraints that govern how we express our thoughts in natural language. This mode of conception yields knowledge that is stable within an individual and largely shared by individuals within a speech community.

10.6 Lexically Expressible Kind Concepts Use Kind-Determined Modes of Explanation and Description

It is generally assumed that lexical concepts such as DOG, TABLE, or TREE are represented in terms of a logical or statistical combination of more primitive concepts, or in terms of more basic causal/explanatory principles that structure concepts within a given domain. These approaches seek to construct kind representations out of kind-independent concepts and principles. So, for example, one may seek to represent the concept DOG out of kind-independent representations FUR, TAIL, BARK, and others. The latter concepts are kind-independent, in that they can be used to construct many different kind representations. FUR and TAIL, for example, may enter into constructing kind representations for both CAT and DOG. This reductionist approach to the representation of lexically expressible kind concepts stands in stark contrast to the theory of conceptual form.

The theory of conceptual form posits kind-determined principles for thinking and talking about each kind of thing. Lexically expressible kind concepts are not built out of some pool of more primitive kind-independent concepts. Nor are they to be understood in terms of some more basic kind-independent explanatory principles. Instead, each kind concept is understood as providing a principle of explanation whereby we understand instances of that kind to have certain properties because they are the kinds of things that they are. So, for example, we understand dogs to have fur, to have tails, and to bark because they are dogs, and we understand cats to have fur, to have tails, and to meow because they are cats. The kind of thing that something is explains why it has certain properties, but not other properties (e.g., we cannot explain why some dogs are brown or lack tails, or why most dogs wear collars, by citing the fact that they are dogs). Although formal explanations of this sort are not very satisfactory from a scientific perspective, they are not tautologous and are explanatory to some degree (Rivera et al., 2023). Furthermore, the properties that are explained via formal explanations are the ones that are understood to be caused by the usually unknown essence by theories of psychological essentialism (Prasada, 2017; Muradoglu et al., 2022). As such, the kind-independent causal mode of explanation assumed by psychological essentialism presupposes the nonreductionist formal mode of understanding that is intrinsic to kind representation.

Lexically expressible kind concepts not only provide kind-determined principles of explanation, for many classes of kinds, they also provide kind-determined principles for quantifying, evaluating, locating, and characterizing things. As discussed in chapters 5 and 7, each kind representation specifies what counts as one instance and/or subkind of that kind, and thus the kind-determined units in terms of which we can quantify that kind. We can, of course, also quantify a kind in terms of kind-independent units such as PILE, BOX, POUND, and LITER, but each kind representation specifies at least one, and possibly two, kind-determined units of quantification.

Furthermore, because kind representations specify the properties that instances of that kind are supposed to have, the kind specifies a standard of evaluation that allows us to evaluate the extent to which instances have the properties they are supposed to have, and thus the extent to which they are perfect. This standard of evaluation is obviously a kind-determined standard.

Kind representations can also specify frames of reference by specifying axes and the orientation of axes that are centered on instances of the kind that allow us to think and talk about where things are in kind-determined ways. For example, it is a kind-determined frame of reference that allows us to specify which side of a table is supposed to face up, and thus determine whether a table is upside down. If the same thing is identified as a hunk of wood, on the other hand, it makes no sense to ask whether it is upside down because WOOD does not determine a reference frame for talking about where things are (Ferenz, 2000; Prasada, 2005). Of course, we can also use kind-independent reference frames, such as allocentric and speaker-determined reference frames, to talk about where things are (Li & Abarbanell, 2019).

Finally, as discussed earlier, kind representations contain a component for representing the character of the kind in terms of the properties to which the kind has principled connections. Kind representations from different classes have principled connections to different types of properties. For example, kind representations for living kinds such as DOG have principled connections to both the parts and stuffs of which they are constituted, whereas kind representations for artifact kinds such as CHAIR have principled connections to the parts of which they are constituted, but not the stuffs of which they are constituted. As such, kind representations in the former class (e.g., DOG) project positions for representing both the parts that instances of the kind are constituted of (e.g., __DOG$_{\text{PART}}$) and the stuffs that instances of the kind are constituted of (e.g., __DOG$_{\text{STUFF}}$). Those projected positions are formally dependent on the kind representation and provide the means for representing properties that characterize the kind in a kind-determined manner—as a dog part (e.g., DOG-TAIL, DOG-SNOUT) or as dog stuff (e.g., DOG-FUR, DOG-BLOOD), and more generally, as aspects of being that kind of thing. Thus, we see that the properties in terms of which we characterize a kind are also represented in a kind-determined manner rather than in kind-independent terms.

As these observations illustrate, lexically expressible kind concepts provide perspectives that allow us to think and talk about the number, location, perfection, and qualities of things in kind-determined ways. One important set of questions raised by these facts concerns how the kind-determined representations are related to their corresponding kind-independent representations. One interesting possibility is that that all nonkind concepts are first represented in kind-dependent terms, and only later in kind-independent terms. If so, that would mean that we start with concepts of kinds of things, and only later form concepts of the features of those things, reversing the standard way that theories of conceptual representation have assumed. Whether or not that turns out to be the case, what is clear is that within the theory of conceptual form, kind concepts are not constructed of more primitive concepts and are themselves primitive, in that they specify their own ways of counting, evaluating, locating, and characterizing things.

10.7 Lexically Expressible Kind Concepts Are Conceptual Primitives

There is no limit, in principle, on the number of new thoughts that we can think. It is usually believed that this capacity requires a mechanism for recursively combining primitive concepts in new ways to think new thoughts (Fodor & Pylyshyn, 1988; Kazanina & Poeppel, 2023; Quilty-Dunn, Porot, & Mandelbaum, 2023). There is no agreement, however, regarding the number, nature, or identity of the primitive concepts that are combined to yield new thoughts.

Notoriously, Jerry Fodor has argued that all roughly word-level concepts (what I've been calling "lexically expressible concepts"), such as DOG, BARN, RED, KILL, and BETWEEN are atomic primitives (e.g., Fodor, 1975, 1998a; Fodor & Lepore, 1998). Such concepts can be combined to think new thoughts like RED BARN IN IDAHO. Fodor provides two negative arguments and one positive argument for the conclusion that lexically expressible concepts are atomic primitives. He argues that because lexically expressible concepts cannot be defined, they must be atomic primitives. Although researchers generally agree that virtually all lexically expressible concepts do not have definitions in terms of a set of necessary and sufficient features, most resist Fodor's conclusion that they are all atomic primitives. He also argues that lexically expressible concepts cannot be prototypes because prototypes do not compose, and any viable theory of concepts must be one, which allow us to compose concepts to think an unlimited number of new thoughts. Fodor notes, for example, that we can use complex concepts such as RED BARN IN IDAHO to think about red barns in Idaho without having any knowledge of what is typical of red barns in Idaho, illustrating that concepts cannot be prototypes (Fodor, 1998a, 1998b). Fodor's positive argument for lexically expressible concepts being atomic primitives is that they contribute exactly what they need to to the complex concept for determining the meaning of complex concepts. Thus, BARN contributes "barn" to the meaning of the complex concept RED BARN IN IDAHO, not "wooden," "smelly," "musty," or anything else, although barns are typically all those things (Fodor, 1998b).

Although researchers agree on little else, they are virtually unanimous in rejecting Fodor's proposal that all lexically expressible concepts are primitives. It is generally assumed that there must be some radically smaller set of primitives. Contra this consensus, the theory of conceptual form provides evidence that lexically expressible kind concepts such as DOG, BOTTLE, and TREE are indeed primitive.

Evidence that lexically expressible kind concepts are primitive comes from comparing them to unarguably complex concepts such as BROWN DOG with respect to how we quantify instances of kinds and how we think and talk about the identity of instances of kinds. Complex concepts such as BROWN DOG do not specify units of quantification. They inherit a unit of quantification from the primitive kind concept that is one of the complex concept's constituents. To quantify instances of complex concepts such as BROWN DOG or RED BARNS IN IDAHO, we use the unit of quantification provided by the kind concept that is a constituent of the complex concept. Lexically expressible kind concepts such as DOG, TABLE, or TREE are primitive kind concepts that do not have any constituent concepts from which they could inherit a unit of quantification. As such, each lexically expressible kind concept specifies its own units of quantification. Each instance is quantitatively equivalent to every other instance of that kind as one instance of that kind. If there is some other way in which each instance of a kind is quantitatively equivalent to every other instance of that kind in terms of some combination of other concepts, it is not known to us. What is known, though, is that each instance of a lexically expressible count kind concept is quantitatively equivalent. As such, lexically expressible kind concepts specify a unit of quantification, and unlike complex concepts, they do not inherit the unit of quantification. If lexically expressible concepts were complex concepts, they too would inherit a unit of quantification from a constituent. That is not possible, however, because lexically expressible kind concepts such as DOG specify instances of the kind as a unit of quantification and that unit of quantification obviously cannot be specified independent of that kind, and thus the unit of quantification could not be inherited from some putatively complex concept of which the kind concept is a constituent.

Similarly, complex concepts such as BROWN DOG do not provide a principle of identity. They inherit the principle of identity from the primitive kind concept that is one of the complex concept's constituents. To ask whether the brown dog in front of us is the same brown dog as the brown dog that we saw earlier is to ask whether it is the same dog. In fact, formulating a question about numerical identity that is framed in terms of the numerical identity of an instance of a complex kind, as we just did, sounds odd. The reason why it sounds odd is because there is not a different form of continuity through time and change that we posit for instances of complex concepts. Complex concepts inherit the principle of identity provided by the kind concept that is one of the constituents of the complex concept. Lexically expressible kind concepts such as DOG, TABLE, or TREE are primitive kind concepts that do not have any constituents from which they could inherit a principle of identity. As such, each lexically expressible kind concept specifies its own principle of identity, which allows us to think that each instance of the kind remains the same instance of that kind throughout its existence.

For example, Fido, as well as every other dog, remains the same dog throughout its existence. If there is some other way in which each dog must remain the same, in terms of some combination of other concepts, it is not known to us. What is known is that each dog must remain the same dog throughout its lifetime. As such, the principle of identity provided by lexically expressible kind concepts is primitive within our conceptual systems, and unlike complex concepts, they don't inherit it from a constituent, as they do not have any.

10.8 Lexically Expressible Kind Concepts Are Primitive, but Not Innate

One reason for the popularity of reductionist approaches to conceptual representation is the general consensus that primitive concepts cannot be acquired and thus must be innate (Laurence & Margolis, 2024).[3] And while more empiricist minded researchers seek to limit the number of innate concepts to a small number of sensorimotor primitives, even more nativist researchers who allow for a larger number of innate and abstract primitive concepts balk at the idea that all lexically expressible kind concepts, including CARBURETOR, are innately given. Instead, it is generally assumed that there must be a limited stock of primitive concepts that are innately given and that all other concepts are constructed out of that limited number of innately specified set of primitive concepts. Jerry Fodor (1975, 1998a) is a well-known exception, who argued that all lexical concepts, not just lexically expressible kind concepts, are primitive in the sense of not having constituent structure and are innate.

The theory of conceptual form departs from both these positions in proposing that all lexically expressible kind concepts are primitive, but they are not innate. To see how that is possible, let us briefly recap the core claims of the theory of conceptual form. According to the theory, there is an innate generative system that yields the constrained variety of perspectives provided by distinct classes of lexically expressible kind concepts. Kind concepts at their core are generative mechanisms that allow us to generate an unlimited number of instance-of-kind representations, as well as subkind representations. In the fragment of the system developed in this book, classes of kind concepts differ along three dimensions. First, classes of kind representations differ in which types of representations they intrinsically generate—instance-of kind and subkind, instance-of-kind but not subkind, subkind, but not instance-of-kind, abstract and concrete instance-of-kind representations. Second, classes of kind representations differ in the types of character that they have because they differ in the number and types of explanatorily significant properties that they have principled connections to and the formally distinct ways in which those types of properties may be realized. Third, classes of kind representations can differ in whether they contain a placeholder for representing an understanding of the kind that is independent of the understanding intrinsic to that provided by the conceptual form of the kind representations within the class.

The first thing to notice about this innate system of knowledge is that it characterizes the conceptual form of kind representations from different *classes* of kind concepts. As such, the innate knowledge system does not characterize any specific lexically expressible concept such as DOG, TABLE, or SAND. Instead, it specifies the formal structure that any such concept may have and thus the perspectives that kinds from different classes of kind representations provide. Lexically expressible concepts are acquired by tokening a concept with one of these conceptual forms by specifying an index to numerically distinguish it from other lexically expressible kind concepts *and* specifying at least some of the properties that the kind is understood to have a principled connection to distinguishing its character from that of other kinds. Neither the index or the properties that characterize a given kind is innately specified. The index can be the name of the kind, or be internally generated when a new kind representation is tokened, and the properties that have a principled connection to the kind and specify the kind's intrinsic character must be acquired via direct experience with one or more instances of the kind or via linguistically conveyed knowledge that ultimately must have derived from experience with instances of the kind in question. As such, lexically expressible kind concepts cannot be innately present, although the forms of principled unity that any given kind concept may display is innately specified.

This account of how lexically expressible kind concepts are acquired via an innate knowledge system that specifies the forms of principled unity that any kind concept must display is analogous to the principles and parameter approach to language acquisition. Universal grammar specifies the formal structure that any human language must display and guides the acquisition of particular languages via specification of open parameters and the acquisition of lexical items. In this approach, no human language is innately specified—what is innately specified is the form and possible parametric variation that any actual human language may display. Similarly, on the approach to conceptual representation developed in this book, no lexically expressible kind concept is innately specified—what are innately specified are the forms of principled unity and possible parametric variation that actually existing lexically expressible kind concepts can display. Which actual languages get tokened depends on the input that the child hears; analogously, the actual lexically expressible concepts that we form depends on the things that we encounter directly or via language.

It is clear then why lexically expressible kind concepts are not innately present. Furthermore, as discussed in the previous section, lexically expressible kind concepts are primitive in the sense that they are not constituted of other lexically expressible concepts and specify their own ways of counting, evaluating, locating, and characterizing things. As such, the innately specified knowledge system for acquiring lexically expressible kind concepts allows us to acquire kind concepts that lack constituent structure and are among the primitive elements that can be combined via syntactic mechanisms to create an unlimited number of complex concepts that possess constituent structure.

The theory of conceptual form is thus in agreement with Fodor's claim that all lexically expressible kind concepts, including CARBURETOR, are primitive concepts in that they lack constituent structure; however, it disagrees with Fodor's and the field's consensus that primitive concepts must be innate because they cannot be acquired.

A key reason that the theory of conceptual form is able to oppose the consensus that primitive concepts cannot be acquired is that the theory hypothesizes that lexically expressible concepts have a formal structure that provides instructions for how to think about the concept's contentful elements. Because concepts have a formal structure that is systematically related to the content of the concept, it is possible for the formal dimension of lexically expressible concepts to be innately given and to guide the acquisition of primitive kind concepts via a process of parameter setting that specifies content that is not innately specified.

In contrast, standard theories of conceptual representation do not acknowledge the existence of conceptual form or assume that the form of a concept is arbitrarily related to its content, and thus the only possibility is for some contentful concepts to be innately given with empiricists and nativists differing mostly about the number and abstractness of those concepts. The theory of conceptual form, on the other hand, allows for the possibility that we come prepared to form concepts that allow us to think about various formally distinct forms of principled unities that kind concepts can display, but that experience determines the specific kind representations we form. There is no limit, in principle, to the number of such primitive concepts we can form.

Although the focus in this book has been on articulating a theory of the conceptual form of kind concepts and it is clear that lexically expressible kind concepts such as DOG, TABLE, and TREE are primitive in the sense that they are not constituted of more primitive concepts, there is good reason to think that the same is true of other lexically expressible concepts. Thus, we can expect that SOFT, GROW, and BETWEEN, along with all other lexicalized concepts, are primitive in the sense that although they possess a conceptual form, just like kind concepts, they lack constituent structure. In this, they all contrast with phrasal concepts such as SOFT SWEATER, GROW SPINACH, and THE DOG IS BETWEEN THE SOFT SWEATER AND THE SPINACH, all of which possess constituent structure (see Pietroski (2018) for a theory of how such complex concepts are interpreted within an internalist framework).

10.9 Domain Specificity Arises from Parametric Variation Allowed by Domain-General Principles

The fact that adults reason about things from different domains in systematically different ways is uncontroversial. Researchers disagree, however, about the source of that domain-specificity. According to some researchers, innate domain-specific principles guide the acquisition of knowledge in different domains from the beginning (e.g., Carey, 1985, 2009; Keil, 1989; Gelman, 1990; Spelke, 1990, 2022), and those domain-specific principles are rooted in specialized areas of the brain (Caramazza & Shelton, 1998). Others understand domain-specificity as arising from learning that different types of properties are more or less important for thinking about things from different domains (e.g., Warrington & McCarthy, 1983, 1987; Warrington & Shallice, 1984; Cree & McRae, 2003) and learning the different patterns of correlations and/or causal structure found for items from different domains (e.g., Devlin, Gonnerman, Andersen, & Seidenberg, 1998; Tyler, Moss, Durrant-Peatfield, & Levy, 2000; Colunga & Smith, 2005; Sloman, Lombrozo, & Malt, 2007).[4]

The theory of conceptual form provides an alternative to both those positions. According to the view developed in this book, kind concepts across domains possess a formal structure that allows us to think and talk about the properties of kinds and their instances and/or subkinds with formally different classes of kinds being distinguished by (1) what they individuate (e.g., instances and subkinds, just instances, just subkinds, or two kinds of instances); (2) the number and types of explanatorily deep properties that they have principled connections to and the ways in which those properties are realized; and (3) whether they contain a placeholder for representing an understanding of the kind that is independent of the understanding intrinsic to that provided by the conceptual form. These formally distinct ways of structuring kind concepts are innate and available for structuring concepts from the beginning. Furthermore, these formal ways of distinguishing classes of kinds correspond closely to distinctions in a commonsense ontology (e.g., living/nonliving, natural objects/artifacts/ living things, and plants/animals/humans (see chapters 5 and 7; see also Prasada, 2017).

The formally distinct classes of kind concepts provide distinct principles for thinking and talking about kinds from distinct content domains. Although those domain-specific principles differ from domain to domain, they are not unrelated to one another on this account. In fact, they are parametric variants that represent the different ways in which things from different domains are intelligible in terms of the domain-general formal mode of explanation, which is implicit in the formal structure of kind representations. For example, although we can explain why a chair has legs by citing the kind of thing that it is, we cannot explain why the chair is constituted of the stuff it is constituted of by citing the fact that it is a chair. The fact that something is a chair constrains, but does not determine, the material that it is constituted of. On the other hand, we can explain both why Fido has four legs, as well as the fact that he is constituted of the types of stuff that he is constituted of (e.g., dog flesh, dog blood, dog hormones) by reference to the fact that he is a dog. This is because within the domain of artifact kinds, we represent a principled connection between the kind of thing that something is and the parts that it is constituted of, but for kinds of living things, we represent a principled connection between the kind of thing something is and both the parts and stuffs that it is constituted of. This difference underlies the different types of properties for which we have explanatory and normative expectations from the different domains. It also underlies the differing patterns of correlations in the two domains, as one can expect there to be strong correlations for characteristics determined by the kind, but weaker correlations for properties that are not determined by the kind and may only be constrained by the kind. Similarly, the differing causal structures that one can expect to find across domains are likely to be aimed at explaining the types of properties that are understood to have a principled connection to kinds in that domain (Prasada, 2017).

The theory of conceptual form thus provides a different way of thinking about the domain-specific patterns within our knowledge of kinds than that provided by existing approaches. One advantage of the theory of conceptual form is that it provides an account of how we can think and talk about how kinds from different domains are similar and different from one another. The domain-specific principles that organize kinds in one domain obviously cannot be used to think and talk about kinds from other domains. The conceptual form of kind concepts provides a domain-general format for representing the domain-specific principles that structure kind representations from different domains, and thus it allows us to align kind representations from different domains and think and talk about both what is distinctive about kind representations from different domains and the ways in which they are similar to one another. If the domain-specific principles organizing kind representations did not share a common format, such thoughts about alignable similarities and differences would not be possible (Gentner & Markman, 1994).

On this account, distinct domain-specific principles are available to guide the acquisition and use of conceptual knowledge from the start without requiring those principles to be rooted in specialized areas of the brain that evolved to support identification and reasoning in different domains. The distinct domain-specific principles arise as distinct parameter settings made available by a domain-general mechanism for thinking and talking about kinds. Whatever evolutionary change made the mechanism for kind representation available also made available the possibility of having domain-specific principles for reasoning about kinds from different domains.

A final difference between accounts of domain-specificity in the literature and the theory of conceptual form is that it can account for differences in the normative expectations that we have for kinds from different domains. This is a natural consequence of the conceptual form of kind representations, but it is not accounted for by existing approaches to kind representation and would require additional assumptions to capture them.

10.10 Conceptual Form and Conceptual Knowledge More Generally

The conceptual forms of kind concepts provide rich perspectives from which to think and talk about kinds, instances of kinds, subkinds of kinds, and their intrinsic properties. The conceptual form of kind concepts specifies the principled ways in which instances and subkinds of a kind are the same, as well as the principled kind-determined ways in which they differ. As such, the theory of conceptual form specifies the principled forms of unity and difference that the human conceptual system comes prepared to think and talk about. The perspectives provided by the conceptual form of a concept are constituted by the principles that each kind concept provides for thinking and talking. In addition to the sortal principles that identify the basic sort of thing that each kind concept allows us to think and talk about, each kind concept also provides a set of principles concerning the character of the kind in terms of the properties that have a principled connection to the kind. Those principles provide starting points for thinking and talking about the kind and its instances and are intrinsically descriptive, normative, and explanatory.

All this knowledge is captured via the formal structure of the kind representation, which provides instructions for how to think and talk about the contentful elements of the concept. Although we may add to conceptual knowledge of this sort and elaborate the principles provided by a given kind concept, we cannot revise that knowledge. As such, this type of conceptual knowledge is stable and is sometimes reflected in the morphosyntactic structure of language. The formal structure of kind concepts also provides the basis for representing knowledge concerning causal or statistical generalization about kinds (see chapter 6). More generally, the knowledge encoded in the conceptual form of kind concepts complements the revisable facts, generalizations, and theories acquired and represented via other means. An interesting possibility is that the formally distinguished classes of kind representations provide the basis for different framework theories. Understanding the scope and limits of each of these formally distinct forms of conceptual knowledge, and how they are recruited in the course of thinking, talking and reasoning, will be an important task for future research.

11 New Horizons for the Study of Conceptual Form

The fragment of the theory of conceptual form developed in this book captures the hidden formal structure of some of our simplest and earliest acquired commonsense concepts. One important task for future research will be to investigate how representations of the proposed sort are accessed and combined with other types of information in real time to think and talk about things. The conceptual form of kind concepts allows us to think and talk about an unlimited number of instances of kinds, most of which we have not encountered and about which we have no direct information. That is possible because the instance-of-kind perspective specifies how to think about things simply on the basis of their being thought to be an instance of a given kind. Of course, the instance-of-kind perspective does *not* specify properties that are not determined by the kind or that do not have a lawful connection to the kind (e.g., the color of a dog). How is information about such properties of instances of a kind incorporated into our thinking about such instances when it is available? Is it incorporated differently than properties of an instance that align with those that are expected on the basis of conceptual form (e.g., being four-legged for a dog)? Or those that contradict the expectation that derives from conceptual form (e.g., being three-legged for a dog)? Does it matter whether the property of the encountered instance has a principled, statistical, or causal connection to the kind? Relatedly, how does our ability to reason about particular objects on the basis of an object's specific causal history, rather than on the basis of the kind of thing it is (see Rips, in press), interact with the perspectives provided by conceptual forms?

Another important task for future research is to investigate the conceptual forms of a broader range of concepts to determine the extent to which the formal elements that enter into the conceptual form of kind concepts may enter into the conceptual forms of other types of concepts. For example, research suggests abstract parallels in our conceptions of objects and events (e.g., Lee & Pappafragou, 2024; Wellwood, Hespos, & Rips, 2018), which raises questions about the extent to which machinery from the fragment of the theory of conceptual form developed in this book can capture those parallels, and how it must be supplemented to represent event concepts or other types of concepts.

The theory in its current form also brings into focus and suggests promising new avenues of investigation. Here, I briefly outline how the theory may (1) guide the investigation of how we think and talk about persons and the groups they belong to, (2) be investigated and extended via computational modeling, (3) suggest some avenues for investigation of how concepts are represented in the brain, and (4) suggest a cognitive perspective on Aristotle's categories and how they may enter into elaborating the theory of conceptual form. My goal is to illustrate how the theory of conceptual form developed thus far facilitates these investigations. How far the ideas mentioned here can be developed and investigated fruitfully is not known, nevertheless, articulating and developing these ideas appear to constitute a worthwhile direction to explore.

11.1 Perspectives for Thinking and Talking About Persons and the Groups They Belong To

Lexically expressible kind concepts such as DOG, TABLE, and HUMAN BEING make available perspectives that specify the characteristics that are true of their instances in virtue of their being the kinds of things that they are. These are determinate characteristics that instances of the kinds are supposed to have. They are shared characteristics of being that kind of thing because they are lawfully connected to being that kind of thing. Instances of the kind are understood to be essentially the same, as they all possess the character that characterizes the kind, and they differ in accidental ways. The concept PERSON provides a radically different perspective.

Each instance of the kind PERSON is expected to *differ* in principled ways from every other instance and similarities among persons is accidental or due to something other than their being persons. So, for example, we expect each person to have their own personality, their own beliefs, values, temperament, and other characteristics. Two persons can have similar personalities, values, or temperaments, but such similarities are accidental from the perspective provided by PERSON. The similarities cannot be explained by their being persons. They are explained via some other factor, such as their being members of the same religion or their having grown up in the same family. As such, instances of PERSON are not merely numerically distinct, with qualitative differences being understood as being accidental. Instead, persons are expected to be both numerically distinct and differ from other persons in *principled* ways. This is because PERSON has principled connections to the *dimensions* on which persons are characterized and distinguished from one another, not on the values that those dimensions can have. Each person generally has their own values on those dimensions. Together, the values of those dimensions specify the individual person's character. The character represents the respect in which we expect the person to differ from other persons, as well as the respect in which we expect a given person to remain the same throughout their existence. As such, PERSON specifies two ways in which we expect a person to remain the same throughout their existence—each person is expected to remain numerically the same person throughout their existence, but they are also expected to have a unique character that distinguishes them from other persons throughout their existence.

In contrast, each instance of HUMAN BEING is expected to be similar to every other instance by virtue of being a human being, and the differences among instances are understood to be accidental or due to something other than their being human beings. The kind concept HUMAN BEING has principled connections to determinate properties that characterize the kind and are supposed to be true of all instances.

Ye (2024) provides some evidence for these differences in the perspectives provided by HUMAN BEING and PERSON. For example, she found that participants judge sentences that state that a characteristic is different for each instance (e.g., "What a person's rights are depends on who he or she is"/"-Every person has their own values") to be true to a greater extent when they are statements about persons than when they are about human beings. Control sentences (e.g., "People can be quite different from one another"/"-Every person changes a lot during the course of a lifetime") showed no difference between the versions with *person* and *human being*. Thus, there are characteristics for which we expect each person to differ to a greater extent than each human being even though every person is a human being. This is possible because PERSON and HUMAN BEING have distinct conceptual forms that provide different instructions for how to think about the things that they allow us to think and talk about.

Furthermore, Ye (2024) provides some evidence that PERSON, but not HUMAN BEING, provides two ways of thinking about the way in which instances are expected to remain the same throughout their existence. She found that statements like "John is not the same person he once was, but is obviously the same human being he has always been" to be judged to be true to a greater extent than statements like "John is not the same human being he once was, but is obviously the same person he has always been."

Another important difference between PERSON and HUMAN BEING (along with the other kind concepts considered in this book) is the fact that each instance of a person is understood to intrinsically belong to a family that is a group of persons of a specific sort. This expands the classes of kind concepts to include kind concepts such as FAMILY, whose instances are groups of people rather than mere instances, subkinds, or ad hoc instances. Other kind concepts whose instances are groups of people include COMMUNITY, ASSOCIATION, and COMPANY, to name a few. Are there formal differences that distinguish different classes of kinds whose instances are groups of people? If so, what are those formal differences?

One dimension on which group-instance kinds differ is whether they provide a principle of individuation. It is straightforward to ask, "How many families are there in this country?" Although there may be some difficulty in determining which notion of "family" we should use, once that has been settled, the query is entirely sensible, and we expect the kind to specify what counts as one instance. On the other hand, the question "How many communities are there in this country?" sounds odd, as there is no such thing as a community that is not some *kind* of community—a religious, linguistic, racial, ethnic, or artistic community, for instance. In such cases, not only does one have to decide which kind of community we want to count, but once that decision has been made, COMMUNITY does not provide a principle of individuation; religious communities are individuated by the religions that have adherents in the country. Similarly, racial communities are individuated by the races represented, ethnic communities by the ethnicities represented, linguistic communities by the languages spoken, and so on. As such, it is the specifying concept that provides the principle of individuation, not the specified concept COMMUNITY. There is a sense in which COMMUNITY is a superordinate concept such as ANIMAL, in that neither has any instances directly. Any individual animal must be some specific kind of animal and cannot be simply an animal of no specific sort (see chapter 5). Similarly, any individual community must be some specific sort of community and cannot be simply a community of no specific sort. On the other hand, FAMILY allows us to think of a family without thinking that it must be some specific sort of family. A group of people can simply be a family without necessarily being some sort of family.

A second dimension along which such "group-instance" kinds differ from one another concerns whether the group intrinsically assigns roles to its members or not. Thus, FAMILY is an example of a kind concept that intrinsically assigns roles to its members, but on the other hand, COMMUNITY is an example of a concept in which members need not possess roles by virtue of being community members.

The introduction of roles into conceptual form brings with it new questions concerning the relation between roles and occupants of roles. Is it possible for a person to change roles within a group? Does that depend on the type of group? What about the type of role? Questions of this sort come to the fore when we begin investigating the conceptual form of PERSON and the groups to which people can belong but have largely been absent from work on conceptual representation in psychology (but see Noyes, Keil, Dunham, & Ritchie, 2023). It appears that attention to the conceptual form of social concepts will reveal rich and intricate perspectives for thinking and talking about persons and the groups of which they are members.

By expanding the study of conceptual form to include social kind concepts, we can hope to have a fuller theory of the forms of principled unity and difference that the human mind comes prepared to think and talk about. We cannot expect to have a theory of all the specific kind concepts that we actually form, as that depends on contingent facts about experience, but we may hope that further investigation of the theory of conceptual form will reveal a theory of the *forms* of principled unity and difference (formally distinct classes of kind concepts) that our minds naturally construct and use in language and thought.

11.2 Concepts as Generative Programs

According to the theory of conceptual form, kind concepts at their core are generative mechanisms of a specific sort. I have sought to elucidate what that sort of mechanism may look like for formally distinct classes of kind concepts and how that informs the ways in which we think and talk about kinds, instances, subkinds, and their properties. The idea that kind concepts are at their core generative mechanisms aligns well with recent work in computational cognitive science, which has sought to model a range of cognitive capacities via generative models of one sort or another.

One approach that seems to be particularly well suited for modeling lexically expressible kind concepts is the Bayesian program learning (BPL) framework, which has been used to model the learning of a large class of visual concepts (Lake, Salakhutdinov, & Tenenbaum, 2015). With that approach, concepts are "simple probabilistic programs—that is, probabilistic generative models expressed as structured procedures in an abstract description language" (Lake et al., 2015, 1333). Their model learns by constructing programs that best explain the observations under a Bayesian criterion and develops hierarchical priors that function as inductive biases that allow previously learned related concepts to help in the learning of new concepts. Lake et al. (2015) show that their model is able to learn a large class of handwritten characters from fifty languages from a single example and can generalize in ways that are similar to the ways in which people generalize. The model builds each character type (e.g., A, B) in terms of a combination of parts, subparts, and relations and how the character may be generated via a set of strokes. In doing so, BPL constructs new concepts (programs) from pieces of previously learned concepts to capture the "causal and compositional properties of real-world generative processes operating at multiple scales" (Lake et al., 2015, 1333).[1]

The model that Lake et al. constructed is remarkable in many respects. Here, I want to highlight some respects in which the BPL framework may be well suited to building computational models of kind concepts as they are understood by the theory of conceptual form, as well as some ways in which the approach would have to be reoriented to capture kind concepts as understood by the theory of conceptual form.

The fact that concepts are represented as programs involving generative models expressed as structured procedures is an appealing characteristic of the BPL framework, as it seems well suited to capturing three characteristics of the conceptual form of kind concepts. First, kind concepts are generative mechanisms that allow us to think and talk about an unlimited number of instances by being capable of generating an unlimited number of instance-of-kind representations. Second, kind concepts provide instructions for how to think and talk about instances and their properties. Third, they capture the type-token distinction by distinguishing the types of characteristics that get combined in different ways to form new types from the types of characteristics that produce variations among the instances. This parallels the distinction between the properties that have a principled connection, and more generally are lawfully connected, to the kind and characterize the kind and the accidental properties that may apply to instances of the kind.

An important limitation of the BPL framework as implemented in Lake et al. (2015) is that it seeks to learn the causal and compositional properties of real-world generative processes only when acquiring concepts. In so doing, the model captures important characteristics of the content of the concept, but not its form. Consequently, the model does not capture the perspective provided by the concept (other than those just mentioned), as the perspective is encoded in the formal structure of the concept, not its content. Furthermore, characteristics of the perspective provided by a kind concept cannot be acquired by constructing programs that best explain the observations as perspectives are not characteristics of the things that the concepts allow us to think about, and thus are not observed.

What is needed is a program that generates different models of the sort that Lake et al. (2015) created for generating kind and instance-of-kind representations. Each generated model would have the formal characteristics that distinguish a class of kind concepts built into it. This would allow different models to specify the perspectives provided by different classes of kind concepts. The contentful elements of kind concepts within a given class may be acquired via procedures of the sort implemented in the Lake et al. model. Another modification that would be needed to implement the theory of conceptual form within a BPL-type framework is that the model needs to be built in such a manner that it can acquire knowledge about the *types* of properties, rather than only acquiring knowledge about how determinate characteristics can be combined and recombined in new ways in new kind representations. Finally, the models must represent *principled* connection to kinds rather than only causal and statistical connections

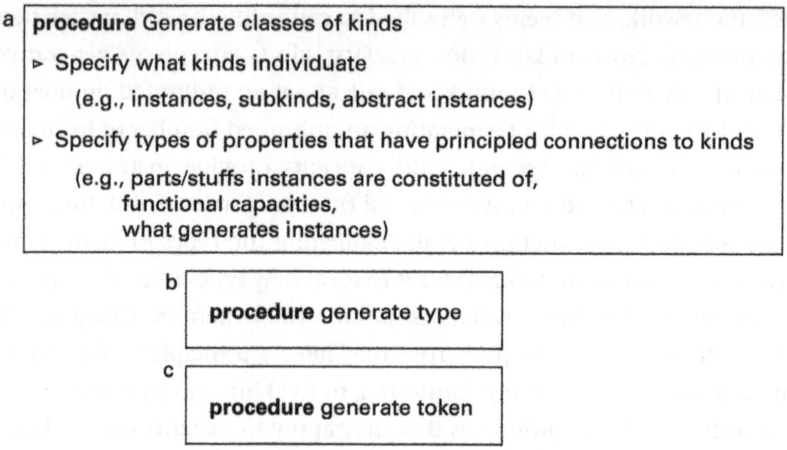

Figure 11.1
Schematic of how the Lake at al. (2015) model would have to be modified to capture the conceptual form of kind concepts. Component (a) is a procedure for generating classes of kind concepts that differ in the types of things that they individuate and the types of properties that they have principled connections to. This procedure then can serve as the input to the type of model used by Lake et al. (2015), which had components for generating types (b) and generating tokens (c). *Source*: Author.

between properties and kinds. As discussed in previous chapters (including chapter 5), kinds concepts from different classes of kind concepts differ in the number and types of properties that they have principled connections to. Figure 11.1 schematically illustrates the addition of these elements via component (a) to the type of model that Lake et al. (2015) used, which included components for generating types (b) and tokens (c).

Modifying the BPL framework along these lines to capture the perspectives provided by lexically expressible kind concepts would provide a powerful tool for investigating the coherence of conceptual form as more elements get added to the theory to accommodate a larger range of formally distinct classes of kind concepts. It would also provide a tool for investigating the types of input conditions that allow learners to acquire the perspectives provided by the various classes of kind concepts. Finally, it would bring specificity to the nature of the computations that are required for using different types of kind concepts to think and talk.

11.3 Investigating Brain Mechanisms Underlying Conceptual Representation

How are concepts represented in the brain? The theory of conceptual form suggests two new ways of approaching that question.

11.3.1 A New Perspective on Domain-Specificity of Conceptual Knowledge in the Brain

It is a remarkable fact that brain damage often leads to selective impairment and preservation of conceptual knowledge from semantically coherent domains (Warrington, 1981; Warrington & Shallice, 1984; Capitani, Laiacona, Mahon, & Caramazza, 2003). For example, there are patients who are impaired in their ability to think and talk about categories of living things, but relatively unaffected in their ability to think and talk about nonliving things. Importantly, there are also patients who show the opposite pattern, ruling out the possibility that concepts from one domain are inherently harder to process, and thus more liable to show impairment when there is brain trauma of some sort. Over the past four decades, researchers have found other patterns of selective impairment for semantically coherent domains such as animals, plants, artifacts, and musical instruments. They have also ruled out potential confounds, which would suggest that these impairments are due to something other than domain-specific organization, and now there is generally agreement that the brain displays domain-specific organization of conceptual knowledge. Furthermore, data from imaging studies with nonpatient populations have generally found convergent evidence regarding the category preferences of the brain regions identified via the study of patients (Martin, 2007; Mahon & Caramazza, 2009). There is disagreement, however, about whether the domain-specificity is *emergent* and the result of domains displaying different patterns of correlations among different types of properties (e.g., Warrington & Shallice, 1984; Tyler & Moss, 2001; Devlin, Gonnerman, Andersen, & Seidenberg, 1998), or whether the brain is *hardwired* to represent and reason about things from different domains in different ways (e.g., Caramazza & Shelton, 1998; Mahon & Caramazza, 2011).

The theory of conceptual form provides a promising characterization of what the nature of the hardwired knowledge that characterizes kind concepts in different domains may be, and it aligns well with the *distributed domain-specific hypothesis* of where the domain-specific constraints on conceptual knowledge representation in the brain come from (Mahon & Caramazza, 2011). As such, the theory of conceptual form suggests interesting new hypotheses concerning how the distributed domain-specific hypothesis may be able to provide new insights into how the brain represents conceptual knowledge.

According to the theory of conceptual form, kind concepts possess a formal structure that encodes the perspective that the concept provides for thinking and talking about kinds, instances, and their properties. That structure is shared by other concepts within the domain and provides instructions for how to think and talk about the contentful elements that characterize the kind and distinguish it from other kinds within the domain. Furthermore, although kind concepts from all domains possess a formal structure that allows us to think and talk about the properties of kinds and their instances and/or subkinds, there are formally distinct classes of kinds that are distinguished by what they individuate (e.g., instances and subkinds, just instances, just subkinds, or two kinds of instances), as well as the number and types of explanatorily deep properties that they have principled connections to and the ways in which those properties are realized. As previously mentioned, classes of kinds that are formally distinguished in this manner correspond closely to domain distinctions in a commonsense ontology (e.g., living/nonliving, natural objects/artifacts/living things, and plants/animals/humans). The differences in the formal structure of kind concepts from different classes of kind concepts provide instructions for thinking and talking about kinds, instances, and the properties that characterize and distinguish kinds within the domain in different ways.

That way of thinking about domain differences fits well with the idea that "domain-specific constraints are aligned with the computational goals required of successful processing of items from different domains" (Mahon, 2023, 222). *Successful processing of kind concepts from formally distinct classes requires different computations.* For example, successful processing of animate kind concepts such as DOG and LION requires different computations from those required for successful processing of artifact kind concepts such as TABLE and FORK (see table 11.1).

The formal structure of animate kind concepts provides instructions for thinking of an instance as one of an unlimited number of instances of that kind; thinking of the distinctness of instances of that kind as being determined by the kind; thinking of the instances as being quantitatively equivalent; thinking of both the parts and substances of which the instances are constituted as being determined by being the kinds of things that they are; and thinking of the instances as having monadic functional capacities, such as the capacity for growth, and dyadic functional capacities, such as the capacity for goal-directed movement and perception because they are the kinds of things that they are. Successful processing of artifact kind concepts such as TABLE or FORK, on the other hand, requires some of the same computations (e.g., that an instance is the same as an unlimited number of other instances of the same kind and the instances are quantitatively equivalent), but also different computations. For example, animate kind concepts determine the stuffs that their instances are constituted of (e.g., dog instances are constituted of dog flesh, dog blood, dog fur, etc.), whereas artifact kind concepts merely constrain but do not determine the stuffs that their instances are constituted of (e.g., table instances can be constituted of wood/metal/plastic, but not paper/cloth/sand).

According to the distributed domain-specific hypothesis, "Domains are individuated by the divergent computational goals that must be solved, not the types of stimuli that drive responses in one or another brain region" (Mahon, 2023, 223). My suggestion is that the formal structure of kind concepts specifies the computations that deploying kind concepts from a given domain involve, and thus the theory suggests systematic ways of investigating where, when, and (hopefully) ultimately how the network of brain regions associated with conceptual knowledge from a given domain may carry out those computations.

The view of concepts developed in this book distinguishes between a lexically expressible concept's form and its content, with the form specifying instructions for how to think about the concept's content—that is, what computations are licensed for that content. Although the formal structures found in different classes of kind concepts are innately given, the contents of specific concepts within each class are acquired via experience. For example, the formal structure of animate kind concepts represents a principled connection between all kinds within the domain and the stuffs of which each kind's instances are constituted, but the character of the stuffs that constitute any specific animate kind concept must be acquired on the basis of experience. Thus, experience with dogs allows us to learn the characteristics of dog flesh and dog fur and how that differs from cat

Table 11.1
Some computations involved in successful processing of kind concepts from different domains

Computation	Animate Kinds (e.g., DOG, LION)	Artifact Kinds (e.g., FORK, TABLE)
Each instance is understood to be quantitatively equivalent.	✓	✓
Subkinds are understood to be merely one way of being that kind of thing.	✓	×
Part identity is determined by the kind.	✓	✓
Substance identity is determined by the kind.	✓	×
Kind determines intrinsic monadic functional capacities.	✓	×
Kind determines intrinsic dyadic functional capacities.	✓	✓
Instances are understood to be produced by instances of some other kind.	×	✓

flesh and cat fur, but the expectation that the kind determined the stuffs of which instances of kinds are constituted is encoded in the formal structure of kind concepts in that domain.

That aligns well with the idea that "connectivity constrains which types of computations are represented in which areas of occipito-temporal cortex. Experience must, of course, provide the content that is represented" (Mahon, 2023, 221). We may thus expect that the connectivity of brain regions that are active during the processing of a kind concept from a given domain may be, in part, specifying the formal structure for concepts within that domain. Whether that is in fact the case remains to be seen, but by clearly distinguishing the form of a concept from its content and specifying how the forms of kind concepts differ from domain to domain, the theory of conceptual form provides a useful way of investigating domain-specific organization of conceptual knowledge in the brain.

Finally, it should be noted that although the distributed domain-specific hypothesis's distinction between form and content of concepts in different domains may capture the distinction between form and content found via the investigation of conceptual form, given our lack of understanding of how anything is stored in the brain (Poeppel & Idsardi, 2022), we cannot rule out the possibility that individual areas of the brain may also be capable of making the relevant distinction between form and content for at least some of the formal characteristics revealed by the investigation of conceptual form. In fact, it may even be that many computational differences may occur intracellularly (Gallistel, 2017, 2021; Akhlaghpour, 2022; Prasada, 2021b). Either way, the theory of conceptual form provides a specification of the types of computations that are relevant to distinguishing concepts from different domains, and thus it may guide the investigation of where, when, and how those computations are carried out in the brain.

11.3.2 A Parts List for Investigating Conceptual Representations in the Brain

Kind representations are fundamental to human cognition, as they provide the means for representing principled unities by representing the principled ways in which an unlimited number of things are thought to be the same, as well as the ways in which they differ. As detailed in chapters 4 and 5, the mechanism for representing kinds has three key components—a descriptive component, an indexical component, and a generative mechanism. All kind representations use all three components with different specifications. As such, the components appear to be good candidates for a parts list for the fundamental representational mechanisms of commonsense conception. It is at this level that Poeppel (2012) and Embick and Poeppel (2005) suggest that we can hope to discover linking hypotheses concerning the brain mechanisms underlying linguistic and cognitive mechanisms.

The descriptive component and the indexical component appear to have analogs outside of the conceptual system, in that indexical representations have been posited to have a central role in perception and visual working memory representations. Specifically, indexical representations are understood as providing the means for binding perceptually distinct characteristics that describe the same object. So, for example, you see a red circle and a blue triangle rather than a blue circle and red triangle because the *descriptive perceptual features* "red" and "circle" are bound to *the same indexical object file* (Kahneman, Treisman, & Gibbs, 1992). Indexical representations are also needed to represent and track qualitatively identical but numerically distinct objects, such as two red circles of the same size (Yu & Lau, 2023). Indexical mechanisms provide a content independent representation that allows us to track an identified object even when its properties change (e.g., Kahneman, Treisman, & Gibbs, 1992; Kibbe & Leslie, 2011; Leslie, Xu, Tremoulet, & Scholl, 1998; Pylyshyn, 1989, 2001; Récanati, 2012; VanMarle & Scholl, 2003; Green & Quilty-Dunn, 2021). In fact, indexical representations allow us to see a change such as a red circle turning into a red triangle, rather than a disappearance of one object (e.g., a red circle) and the appearance of another object (e.g., a red triangle) in a single place. A change involves a single index to which different descriptive features are bound at different times. On the other hand, a change is not perceived if two indices are deployed; in that case, we see a disappearance and an appearance.

There is a growing body of data concerning the neural underpinnings of the indexical mechanism involved in visual perception and visual working memory (e.g., Naughtin, Mattingley, & Dux, 2016; Xu & Chun, 2006; Zhang, Zhang, & Famg, 2020; Zhu, Zhang, & von der Heydt, 2020); however, Yu and Lau (2023, 4) note that "a broad, field-wide effort is needed to develop and refine neural markers for indexicals." Yu and Lau (2023) also suggest that conceptual indexical representations may be needed to bind nonperceptual properties of objects, as well as to think and talk of things as tokens of types.

Instance-of-kind representations of kinds such as DOG and TABLE, whose instances are visually perceptible, have been the focus of much of this book and are a good place to begin investigating the nature of conceptual indices. Identifying the ways in which conceptual indices are similar to, and different from, their perceptual analogs could help shed light on their formal and computational differences and the brain mechanisms underlying them.

A key similarity is that instance-of-kind representations, like perceptual indices, can have properties attached to them that can be used to distinguish among different instances of the kind. Importantly, the identities of the instance-of-kind representation, as well as of the perceptual index, are not dependent on those properties. In fact, for instance-of-kind representations, we often do not know what those distinguishing properties are. So, for example, we may have attached the property of being brown to the instance-of-kind representation of one dog (e.g., DOG_{13}) and the property of being black to a different instance-of-kind representation (e.g., DOG_{47}), and thus be able to distinguish the two dogs via those accidental properties (e.g., the dog that is brown is hungry, not the dog that is black); however, they need not be distinguished in that way. In this instance, DOG_{13} remains the same dog even if it ceases to be brown, and the two dogs remain distinct dogs even if they are no longer distinguishable on the basis of their color (or any other dimension). These considerations suggest that instance-of-kind representations may be formally just like perceptual indices and thus may be realized via similar brain mechanisms.

Key differences between perceptual indices and instance-of-kind representations include the fact that the latter can be used to think and talk about things with which we have no direct perceptual contact, including things that do not exist. Such nonexistent objects can be instances of kinds that actually exist (e.g., a dog that is just like Fido that does not actually exist in reality), or instances of imaginary kinds such as unicorns. Perceptual indices can be used to track only actually existing objects that we have perceptual access to. Furthermore, perceptual indices are limited in number. Conceptual indices are not. We can, in principle, think and talk about an unlimited number of instances of a kind. Finally, perceptual indices are untyped—they identify an object without specifying the type of object it is or describing the object in a fixed manner. Conceptual indices are typed. They are indexical mechanisms that identify instances of a kind in a content-independent manner, which allows us to think and talk about an identified instance-of-kind (e.g., DOG_{13}) even when the instance-of-kind's properties change or we have no information about the specific instance's characteristics. As such, the mechanism allows us to identify and think about something as an instance of a given kind (e.g., as DOG_{13} rather than simply as an object of an unidentified type). These differences suggest that conceptual indices cannot be simply long-term-memory versions of the indexical representations operative in perception and visual working memory.

Both the similarities and differences between perceptual indices and instance-of-kind representations are accounted for if an instance-of-kind representations takes a form that is similar to that of a perceptual index but is produced by a generative mechanism for representing a specific kind of thing, and thus has the capacity to generate an unlimited number of instance-of-kind-representations. The mechanism for representing the kind is stored in long-term memory and is not itself merely an indexical mechanism. It also has a component that characterizes the kind and distinguishes it from other kinds in terms of intrinsic descriptive properties that have a principled connection to the kind. Furthermore, the mechanism for representing the kind takes a form whereby it can generate an unlimited number of instance-of-kind mechanisms. The index can be the name of the kind, as it represents the kind without describing it. The kind can thus be identified via this index (its name) and it could also be identified via the properties that have a principled connection to the kind and thus are bound to that index in long-term memory.

Throughout this book, I have described kind representations simply as generative mechanisms that have the capacity to generate an unlimited number of instance-of-kind representations, with the further constraint that each of them is characterized and distinguished from other kind representations both indexically and via their intrinsic character, which can be specified in terms of the properties that have a principled connection to the kind. That level of specificity has been sufficient for capturing the broad range of linguistic and conceptual phenomena discussed in the book. If, however, we are interested in investigating how the mechanism is related to perceptual indices, the mechanisms of syntax, or how it is realized in the brain, it will be useful to try to specify kind mechanisms in further detail. It is with these goals in mind that I suggest the following way of characterizing the mechanism for kind representation.

11.3.3 Specifying the Generative Component of Kind Representations

Statements about kinds of things are understood to be about an unlimited number of instances of that kind (Fenton & Prasada, 2024). Furthermore, kind representations allow us to think and talk about an unlimited number of things as instances of that kind by generating an unlimited number of instance-of-kind representations. As such, each kind representation is capable of generating a discrete infinity of instance-of-kind representations. In this, kind representations are similar to our linguistic capacity to generate a discrete infinity of hierarchically structured sentences and our capacity to generate a discrete infinity of natural numbers. All three capacities appear to be uniquely human and require a mechanism for generating a discrete infinity of mental representations given finite resources. Ideally, all three can use the same mechanism.

Syntacticians have posited a generative structure building mechanism called *Merge,* which has the capacity to characterize a discrete infinity of sentence structures. Merge is a recursive mechanism that takes two syntactic objects and returns a new syntactic object that can be combined with another syntactic object without limit (Chomsky et al., 2023). So, for example, Merge applied to the syntactic objects *the* and *child* (e.g., Merge (the, child)) yields a new syntactic object (e.g., {the child}), which can be merged with another syntactic object to create another new syntactic object, and so on, with no limits in principle. Although Merge is understood to be binary in its operation in the syntax (Chomsky et al., 2023), it can be applied to a single element in other contexts. Chomsky (2008) notes that Merge can yield the successor function and thus generate a discrete infinity of numbers via its application to a single element: "Suppose that a language has the simplest possible lexicon: just one LI, call it 'one.' Application of Merge to the LI yields {one}, call it 'two' Application of Merge to {one} yields {one, {one}}, call it 'three.' And so on. In effect, Merge applied in this manner yields the successor function" (Chomsky, 2008, 139).

An interesting possibility is that Merge may also be used to represent kinds and the unlimited number of instance-of-kind representations that are contained in any kind. If, for example, Merge operates on a single indexical element, which can be identified via the form which is pronounced /dog/ and which has the properties a_1, a_2, a_3, . . . bound to it (e.g., /dog/—< a_1, a_2, a_3 . . .), and we also assume that Merge operates on the indexical element, but not the properties that are bound to it, it will yield {/dog/}, which may be understood to be an instance-of-kind representation (e.g., DOG$_1$). Applied again, it will yield {/dog/, {/dog/}, which may be understood to be an instance-of-kind representation and another instance-of-kind representation (e.g., DOG$_1$ and DOG$_2$). Applied yet again, it will yield {/dog/, {/dog/, {/dog/}}, which may be understood to be an instance-of-kind representation, another instance-of-kind representation, and yet another instance-of-kind representation (e.g., DOG$_1$, DOG$_2$, and DOG$_3$), and so on . . .

Thus, we see that Merge (/dog/—< a_1, a_2, a_3 . . .) can generate an unlimited number of instance-of-kind representations. Each of those instance-of-kind representations is an indexical representation that allows us to think and talk about a single instance of that kind. That instance is expected to have the properties that characterize the kind and distinguish it from other kinds (a_1, a_2, a_3 . . .), but it cannot be known whether a given instance actually has any of those properties without examining the instance itself or receiving information about the instance from another source.

Characterizing the generative mechanism for thinking and talking about kinds in this manner has a number of advantages. First, it explicitly distinguishes the descriptive, indexical and generative components of kind representations, suggesting that those are members of a plausible "parts-list" for conceptual representation for which we may seek to develop linking hypotheses (Poeppel, 2012). Second, it allows us to see how the generative kind mechanism and the instance-of-kind representations that it generates build on perceptual indices, and thus suggests how investigating the brain bases of perceptual indices may inform our understanding of kind and instance-of-kind representations. Similarly, it suggests that comparisons with perceptual indices may help elucidate the brain mechanisms involved in realizing the generative component of kind representations. Third, it uses a generative mechanism that also appears to be operative in two other distinctively human cognitive capacities, potentially making questions concerning the evolution of these aspects of human cognition more tractable.

Although one might wonder whether the generativity of numbers and of syntax likely have a common source, the generativity of lexically expressible kind concepts and of syntax very likely do have a common source, as the latter provides the means for combining the former to create linguistically expressible thoughts. As has been detailed throughout this book, lexically expressible kind concepts provide abstract perspectives for thinking and talking. Those perspectives appear to be absent from other nonlinguistic mental representations that one might call "concepts" in humans or other species. Those representations may have similar content as lexically expressible concepts, but, as far as we know, they lack the formal structure that specifies the perspectives provided by the lexically expressible kind concepts with similar content. Theories of the evolution of language must account not only for the mechanisms that allow the combinatorial properties of linguistic expressions, but also for the perspectival properties of the conceptual elements that are combined to produce linguistically expressible thoughts, as we do not have human language without having both. As such, it would be helpful if the source of generativity in syntax and lexically expressible kind concepts is one and the same.

It remains to be seen whether a parts list that identifies components of the sort described previously can be used to formulate and investigate sensible bridging hypotheses. However, unlike existing approaches to conceptual representation, the theory of conceptual form suggests a plausible starting point for such investigations.

11.4 A Cognitive Perspective on Aristotle's Categories

As we have seen, to think and talk about Fido and Jett, we use the concept DOG, which is at its core a generative mechanism that can generate an unlimited number of instance-of-kind representations. Those instance-of-kind representations are numerically distinguished via an index that has no intrinsic descriptive content. This allows us to think of Fido and Jett as being distinct dogs even if they happen to be qualitatively identical. Furthermore, each of those dogs is understood to be exactly one dog, and thus quantitatively equivalent qua dogs. Although Fido and Jett are different instances of the kind "dog," they are the same kind of thing (dogs). On the other hand, Fido and Jett are different kinds of things than Fluffy the cat. Being different kinds of things, rather than merely different instances of a given kind, Jett and Fluffy cannot be merely numerically distinct; they must also differ in the character that they have by virtue of being the kinds of things that they are. Jett is expected to have the qualities that dogs have by virtue of being dogs, and Fluffy is expected to have the qualities that cats have by virtue of being cats.

Thus, we see that to think and talk about Jett, Fido, or Fluffy, which Aristotle would call *substances*, we must have access to representations of the kinds that they belong to, what Aristotle called *secondary substances*, as well as representations of *quantity* and *quality*—two more of Aristotle's categories. We cannot simply talk about a substance (e.g., Jett) without identifying the substance as an instance of a kind, which in turn requires us to distinguish the instance from other instances in a way that does not require qualitative difference but does require quantitative equivalence. Furthermore, thinking of substances of different kinds, such as Jett and Fluffy, requires thinking of them as being qualitatively distinct in various ways because they are the kinds of things that they are.

Within the theory of conceptual form, the intelligibility of Aristotelian substances is intrinsically intertwined with secondary substances, quantities, and qualities. The instance-of-kind perspective, which allows us to identify Jett as a dog or Fluffy as a cat, requires conceptual representations such as DOG and CAT that have the formal structure outlined in chapter 4 and illustrated in figure 4.4. That formal structure embodies the perspective provided by the concept for thinking about things and their properties by providing instructions for how to think about the kind, its instances, its subkinds, and the contentful properties of the things that the concept allows us to think and talk about. Those instructions relate substances, secondary substances, quantities, and qualities in a manner that is intrinsic to the perspective provided by that concept (as outlined earlier). In this context, the Aristotelian categories identify necessary components of the mechanism in terms of which substances such as Jett and Fluffy are intelligible to us. As such, the Aristotelian categories are components of the lenses through which we think about things, and thus, as Aristotle and Brentano (1982) suggest, we come to know about the lenses and their properties only incidentally or by the way.

An interesting question for future investigation is whether more detailed investigation of the conceptual form of kind concepts will reveal the other Aristotelian categories as further necessary components of the conceptual form of at least some classes of kind concepts. Aristotle named different numbers of categories in various works, but the most expansive listing includes ten categories. Studtmann (2012, 66) provides the following list and examples of Aristotle's categories:

> (1) a substance, like a man; (2) a quantity, like a line two cubits long; (3) a quality, like the white; (4) a relation, like the double; (5) somewhere, like in the Lyceum; (6) at some time, like yesterday; (7) being in a position, like lies; (8) having, like is shod; (9) acting, like cuts; or (10) being acted upon, like is cut. (*Cat.* 1b25–2a4)

In this book, I have described all nonquantitative properties that are aspects of a kind's character as qualities. The only further characterization of the component that characterizes and individuates kinds that I provided was that the component includes properties that are relevant to Aristotle's modes of explanation. That characterization of the component that represents the character of a kind is certainly too crude. Perhaps closer investigation will show that a better theory of conceptual form will result if we articulate the component that characterizes kinds in terms of the categories and divisions of those categories that Aristotle proposed.

While the value of that project remains to be seen, it is likely that at least two of the remaining categories—place and time—may have a special role within the conceptual form of kind concepts. First, like other properties, they can characterize certain kinds. For example, certain kinds have principled connections to *where* their instances belong (e.g., sharks, by virtue of being sharks, live in the ocean), as well as *how long* instances of the kind are supposed to live (exist) (e.g., sharks, by virtue of being sharks, live twenty to thirty years.). Second, they enter into characterizing how we think and talk about the relation between instances of object kind concepts such as DOG, TABLE, and TREE. Specifically, they are needed to capture the intuition that two instances cannot be at the same place at the same time. Third, the place that characterizes where instances of a given kind are supposed to be can enter into explaining why they have certain characteristics. For example, sharks may be understood to have fins *because* they live in the ocean. Similarly, the place that a given group has within a society may help explain why they have certain characteristics via so-called structural explanations (Vasilyeva & Lombrozo, 2020). Finally, there are things like houses, boxes, and bottles that are intrinsically conceived of as both a kind of object and a kind of place. So, for example, "John ran into the house" could mean that he collided with an object or that he entered a place. This suggests that our commonsense conception of certain Aristotelian substances (e.g., house) requires us to simultaneously identify the substance as an instance of a kind of thing and as a place. In such cases, place does not merely characterize substances; it enters into identifying what that substance is.

Within the theory of conceptual form, the categories are not merely a catalog of the highest forms of different forms of existence. The theory of conceptual form suggests that at least some Aristotelian categories identify necessary components of the formal mechanisms in terms of which we think and talk about the things that we encounter. As such, the categories appear to be necessary components of the mechanisms that represent our commonsense conception of what things are and why they have the properties that they do. Furthermore, as discussed earlier, the conceptual form of kind concepts yields principles for thinking and talking about things insofar as they are the kinds of things that they are.

In previous work, I have suggested that these principles display the characteristics that Aristotle thinks scientific knowledge should have (Prasada, 2021a). While it is clear that the Aristotelian categories and the principles embodied in the conceptual form of kind concepts cannot provide the understanding that modern science seeks, they may nevertheless be intrinsic to the form of understanding provided by commonsense conception. As such, further investigation of how the Aristotelian categories enter into our commonsense conceptions is likely to deepen or understanding of the abstract and intricate perspectives in terms of which we think and speak about the things around us.

12 A Bird's-Eye Perspective on Conceptual Form

What, then, is conceptual form? And what is it good for? Answers to those questions are spread throughout the book, along with the evidence, justification, and arguments for those answers. Here, I distill the key characteristics of conceptual form and the advantages that it affords our conceptual systems.

12.1 What Is Conceptual Form?

- A concept directs our attention to the concept's content. But a concept also specifies certain *perspectives* or *ways* of thinking about the content. Perspectives are encoded in the concept's formal structure, its conceptual form, which provides instructions for how to think and talk about the things that the concept allows us to think and talk about.
- Our most basic concepts (e.g., DOG, BOTTLE) allow us to think of things as instances of a basic-level kind, as well as about the kinds themselves. Those concepts have three types of components: (1) a generative component that captures the nonaccidental unity of the generated instance-of-kind representations; (2) a component that represents the principled qualitative unity that characterizes the kind and distinguishes it from other kinds, is expected to be present in instances of the kind, and is realized in systematically distinct ways by subkinds of the kind; and (3) a typed index, which represents instance-of-kind representations within a kind as numerically distinct.
- The conceptual forms of all kind concepts are constituted from these formal components.

- The qualitative character of a kind is represented via properties that have a *principled connection* to the kind. Principled connections intrinsically have an explanatory, normative, and statistical dimension and differ from statistical, causal, and logical connections found in previous research on conceptual representation. Properties that have a principled connection to a kind are understood to be properties that the kind has by virtue of being the kind of thing that it is and are represented as aspects of being that kind of thing. Furthermore, they are properties that instances of the kind are *supposed* to have by virtue of being the kinds of things that they are. Instances that lack them are understood to be defective or incomplete. Finally, properties that have a principled connection to a kind are expected to be generally present in instances of that kind.
- Principled connections are intrinsic to the conceptual form of kind representations. All kind representations are characterized and distinguished from other kind representations in terms of the properties that they have principled connections to. Kinds can *additionally* be characterized in terms of properties that have lawful causal or statistical connections to the kind, but they need not be.
- Different combinations of the components of kind representations and how those components are realized yield formally distinct classes of kind representations. Those formally distinguished classes correspond to commonsense ontological distinctions.
- The formal structure of kind representations affords formal explanations of properties that characterize kinds in all classes of kind representations. The formally distinguished classes differ, however, in the other modes of explanation (e.g., causal, teleological) that they afford for explaining the properties that characterize the kind. As such, the formally distinguished classes capture the distinct ways in which kinds from each of those domains are known to us.
- Classes of kind representations are also distinguished by whether their kind representations contain a placeholder for representing an understanding of the kinds as they are in and of themselves (independent of how they are known to us). The placeholder may be filled with what we take to be the essence of the kind, but it may also be filled by truly scientific theories. As such, the conceptual form of kind concepts represents things as they are known by us, but it also supports the possibility

for developing an understanding of what things are in themselves for certain classes of kinds.

- Instance-of-kind representations carry information about the mode of existence of an instance as either actually or potentially existing. Actually and potentially existing instances need not differ in their qualitative characteristics, and thus the actual existence of an instance is marked via an untyped index with no intrinsic content. Kinds are understood to exist actually or not at all, and thus kind representations have an untyped index, which marks the kind as actually existing and directs attention to the kind.
- Lexically expressible kind concepts are atomic and provide *principles* for thinking and talking about the concept's content. Those principles specify kind-dependent ways of quantifying, existing through time, characterizing, evaluating, locating and orienting things. Lexically expressible kind concepts can also be characterized by nonaccidental causal and statistical generalizations.
- In contrast, concepts that are expressed via phrasal expressions involve conceptual combination and thus cannot be assumed to have an intrinsic unity that provides principles for thinking and talking about their content. They can, however, be characterized by nonaccidental causal and statistical generalizations.
- Lexically expressible kind concepts are atomic but not isolated. Although lexically expressible concepts are not constituted of other concepts, they have a formal structure that intrinsically affords connections to concepts of certain types (concepts of the types of properties that they have principled connections to).

12.2 What Is Conceptual Form Good For?

- The conceptual form of kind concepts allows us to think and talk about kinds, which are abstract entities we can never encounter, as well as an unlimited number of instances of those kinds, even though we usually encounter only a very small fraction of those instances and have very little specific knowledge about those instances that we do encounter.

- Because the conceptual form of kind representations from different content domains are constituted by the presence/absence of the same set of formal components, the conceptual form of kind concepts provides a uniform format in which kind concepts from different ontological domains can be aligned and compared.
- Because there is a distinction between a concept's form and content, domain differences between how we think and talk about things can be represented without having to represent superordinate concepts like ANIMATE BEING or ARTIFACT. For example, we think and talk about cats and dogs in similar ways because the conceptual form of DOG and CAT instructs us to think of their content in similar ways, and dogs and tables in different ways because the conceptual form of DOG and TABLE instructs us to think of their content in different ways.
- Because there is a distinction between a concept's form and content, we may innately possess knowledge of the formally distinct classes of concepts that we can form and have that knowledge guide the formation of concepts. This allows for a view in which there are no innately given concepts, only innate constraints on the formal structures of concepts.
- Because the conceptual form of kind concepts is unrevisable, it is stable and may be reflected in the morphosyntactic structure of language.
- The theory allows for the coexistence of intuitive understandings and scientific knowledge, even when they are in conflict, because the conceptual form of kind concepts distinguishes between the perspective provided by the concept that captures the understanding of the kind as it is known to us and a placeholder for the scientific understanding of the kind as it is in itself,.
- Conceptual form helps explain many linguistic phenomena, including those listed in table 1.1, how generic knowledge is expressed, the systematic polysemy of kind terms, and the interpretation of metaphors.

That is a bird's-eye perspective of what conceptual form is and some of what it is good for. The fragment of the theory of conceptual form developed in this book sheds light on a wide range of conceptual and linguistic phenomena. I hope that this will inspire continued investigation of conceptual form, as it is likely to yield insights into the nature of concepts and how they are represented, acquired, processed, and realized in the brain.

Notes

1 Situating the Study of Conceptual Form

1. In both psychology and philosophy, there are many different views on what kinds are, whether kinds exist, and what kinds of kinds there may be. For our purposes, what matters is that generic interpretations of noun phrases are understood to be about kinds of things (e.g., "The lightbulb was invented by Edison"), whereas nongeneric interpretations of those noun phrases are about an instance or some set of instances of that kind (e.g., "The lightbulb was next to Edison"). Given that, we need to discover the formal properties of concepts like LIGHTBULB that allow us to think and talk about both the kind and its instances. That is exactly what I seek to do in chapters 3–5.

2. The exemplar approach is regularly presented as a theory of conceptual representation; however, Murphy (2016) argues (convincingly, to my mind) that whereas there are clearly exemplar effects, there is no exemplar theory of concepts.

3. A notable exception is the idea that concepts are atomic and lack constituent structure (e.g., Fodor, 1998a). That approach, like other existing approaches, does not provide an account of the perspectival nature of lexically expressible concepts. For a discussion of the sense in which lexically expressible kind concepts may be atomic *and* have conceptual form, and thus be perspectival, see chapters 5 and 10.

4. Although people believe that the essence really exists in members of the category, they do not come to that belief because they have observed the essence in members of the category. Rather, the belief is part of the perspective provided by the structure of the framework theory.

2 The Perspectival Nature of Commonsense Concepts

1. We can, of course, consider two forks or lions to differ in principled ways if we think of them as being different kinds of forks or kinds of lions. However, we do not have to think of two lions as being different kinds of lions (e.g., Barbary lions and

Cape lions) in order to think that they differ in principled ways—they can simply be a male and female lion.

2. Figure 2.1 summarizes only some of the perspectives provided by the classes that the four chosen concepts belong to. Not pictured here are differences in the perspectives that different classes make available for thinking about subkinds and how subkinds are related to kinds. We will see more about those differences in chapters 5 and 7.

3. I am using "representation" in the manner that it is used in linguistic theory and computational theories generally, in that representations are what computations are performed on. The question of how perspectives are represented is about the form of the mental symbols that enter into computations that express perspectives. Representations in this sense do *not* require some mind-external thing that they stand in some relation to by virtue of which they are representations (see Chomsky, 2003).

4. This proposal aligns with internalist approaches to semantics more generally (e.g., Chomsky, 2000; Knowlton et al., 2021; Murphy, 2023).

5. Of course, we need not think of the thing and its properties in this manner if we think about the thing from a scientific perspective rather than via the perspective provided by a commonsense concept.

3 The Instance-of-Kind Perspective and Limits of Existing Approaches

1. This remains true even if our philosophical or scientific theorizing suggests that things in the world can be unspecified with respect to their kind.

2. An exception is the exemplar approach, which does not form summary representations. Instead, it merely represents some examples of categories. Murphy (2016) convincingly argues that whereas there are undeniably exemplar effects in research on concepts, there is no coherent exemplar theory of concepts. The exemplar approach cannot represent the instance-of-kind perspective, as it has no way of representing kinds as distinct from its examples. As a consequence, I will not be discussing this approach further.

3. For the current purposes, it will be fine to leave the nature of this combination unspecified.

4 Mechanisms for Thinking About Kinds and Instances of Kinds

1. Another reason to resist the conclusion that kinds can exist potentially in the same sense that instances of kinds can exist potentially is that a potentially existing instance of a kind is the same kind of thing as an actually existing instance of a kind. This cannot be the case for kinds. In what sense is a kind that does not

exist at a given time (e.g., computers in the nineteenth century) the same kind of thing as actually existing kinds at that time (e.g., dogs, tables, and trees)? Not yet existing kinds (putatively potentially existing kinds) bear no positive formal relation to already actually existing kinds, they are simply different. In contrast, potentially existing instances of kinds bear a *same-kind* relation to actually existing instances of kinds and differ only numerically and in their mode of existence.

2. To help keep track of whether the representation is of a kind or an instance of a kind, the actual mode of existence of the kind is represented via an untyped Greek-letter index that is a *superscript*, whereas the actual mode of existence of an instance-of- kind representation is represented via an untyped Greek-letter index that is a *subscript*.

3. In passing, I note the oddness of this formulation. Asking how long Fido has been/was/will be hungry/a puppy/a pet/a nuisance all sound reasonable, but asking how long Fido has been/was/will be a dog sounds odd. The obvious answer seems to be that Fido will be a dog so long as he exists, however long that is . . . *unless* something extraordinary happens and he is transformed into something else.

4. One may say that what remains the same is the stuff or the matter (but not any specific kind of matter), but we have no independent and positive characterization of stuff or matter. We can, at best and derivatively, say that stuff is that which is capable of being either the matter of a pumpkin or the matter of a carriage.

5. Although the mechanism allows for the possibility that we can think of an instance of one kind turning into an instance of another kind in ways that are mysterious, it does not *require* such transformations to be mysterious if we have means for identifying what remains the same or have access to knowledge of the types of causal forces that would effect a given transformation (e.g., the transformation of a coffeepot into a watering can is very intelligible).

6. I discuss the claim that specification of the conceptual form of kind concepts results in principles rather than mere facts or generalizations in chapters 6 and 10. Here, I note simply that specification of (13) yields a representation that is intrinsically general, normative, and understood to be in actual existence. Furthermore, counterexamples do not invalidate the principles. These are prima facie characteristics of principles.

7. In both philosophy and psychology, it is standard to assume that our conceptual knowledge is always open to revision (e.g., Putnam, 1975; Carey, 2009; Strevens, 2012). I take as an empirical question whether any given piece of knowledge is actually revisable, even if arguments can be presented for why it should always be possible to revise one's knowledge. In chapter 10, I discuss some evidence suggesting that the principles provided via a specification of conceptual form are not revisable.

5 Generating Perspectives

1. Note that although existing approaches to conceptual representation and the Uniform Conceptual Form hypothesis are similar in this manner, they differ in that the Uniform Conceptual Form hypothesis specifies a single perspective from which we think and talk about all kinds of things, whereas the existing approaches are nonperspectival and represent the structure of the content of the concepts only.

2. It is interesting to note that it is very difficult to get the subkind reading without a lot of stage setting. For example, "I like many kinds of books: picture books, textbooks, reference books . . . but my favorite book is the art book" seems to work but is not great. The subkind reading is always not preferred, but it appears to be even less so for kinds like "book." It is an interesting question for future research to determine why this is so.

3. The one-many relation between an abstract instance and the concrete instances related to it are often said to involve a type-token relation; however, we do not think and talk of abstract instances of books (e.g., *Midnight's Children*) as a type of book, but rather as a single book (more on this point later in this chapter). Art books, reference books, hardcover books, and electronic books are all examples of types of books.

4. The exception is concrete instances of BOOK that are anthologies, which contain the intelligible matter of multiple abstract instances of the kind "book." It is worth noting that such exceptions are understood to not be accidental, but to be true of a specific subkind of books—ANTHOLOGY.

5. It is possible that there may be a more general way of characterizing dual character concepts such that abstract values provide only one way for us to provide an explanatorily "deeper" characterization of a kind (Knobe, under review).

6. Phillips and Plunkett (2023) argue that there may not be a class of dual-character concepts, on the basis of the finding that concepts that are not dual-character concepts can behave like dual character concepts given the right contextual support. But concepts like SCIENTIST and PHARMACIST differ in that the former, but not the latter, support dual categorization judgments in the *absence* of any context (see experiment 3 of Knobe et al., 2013). That is possible because concepts such as SCIENTIST are intrinsically characterized in terms of both abstract values and concrete properties, but concepts such as PHARMACIST are not. The fact that there can be circumstances in which an additional ad hoc characterization of concepts like PHARMACIST is possible is interesting, but it does not erase the difference between the intrinsic formal structures of concepts such as SCIENTIST and PHARMACIST.

7. The proposal here is analogous to the relation between the form and meaning of verbs hypothesized within syntactic bootstrapping theories (e.g., Gleitman, 1990).

Verb classes are formally distinguished from one another by the number and types of arguments they take. Analogously, it is being proposed that types of kinds are distinguished from one another by the numbers and types of explanatorily significant properties that they have principled connections to. Second, verb classes distinguished by the number and types of arguments they take are understood to be mapped onto distinct types of relational meanings. Analogously, kind representations that are formally distinguished by the number and types of explanatorily significant properties that they have principled connections to are mapped onto kinds that have distinct types of explanatory structure.

8. Although it has been previously noted that Aristotle's modes of explanation apply differentially to concepts and lexical items for kinds from distinct domains (e.g., Moravcsik, 1975, 1981, 1998; Keil, 1994; Pustejovsky, 1995), the question of how the explanatorily significant property is related to the kind is not usually explicitly considered. The present proposal is that there is a *principled connection* (rather than a logically necessary or probable or causal connection that is the type of connections that are usually being considered) between the kind and the explanatorily significant property. As previously discussed, principled connections intrinsically support formal explanations of properties that have a principled connection of a kind, as well as supporting normative expectations for those properties (instances are supposed to have those properties); in addition, they are expected to generally exist in instances of the kind. See chapter 6 for a discussion of a number of linguistic and nonlinguistic phenomena that depend on whether a kind is characterized by a property with a principled, causal, or statistical connection to it.

9. The distinctions between different classes of kinds implicated by differences in the number and types of explanatorily significant properties is not independent of the ways of distinguishing classes of kinds in terms of the forms and types of things that kinds individuate. An important topic for future research is to investigate the ways in which these two ways of distinguishing classes of kinds are independent and what constraints there may be on the classes of kinds that may be derived via possible combinations of the two ways of distinguishing classes of kinds.

10. The literature doesn't generally distinguish between ad hoc subkinds and ad hoc kinds. We distinguish them here because, as will become evident, they take different conceptual forms.

11. The theory that I develop is sensitive to the lexical/phrasal distinction. So it treats the concepts underlying "bird" and "blackbird" as being similar to and different from the concept underlying "black bird" and "birds that are black." Both BIRD and BLACKBIRD are understood to be kind representations with characters that distinguish them from other kinds and are specified by the properties that the kinds have principled connections to. In contrast, BLACK BIRD and BIRDS THAT ARE BLACK are not kind representations, so they do not have a character that distinguishes them

from other kind representations and other phrasal concepts. They are distinguished from one another by their constituent structures. Furthermore, BLACK BIRD and BIRDS THAT ARE BLACK are understood to provide the means to think about birds of any kind that are black. As there may be multiple kinds of birds that are black (e.g., crows, grackles, ravens), the category lacks an intrinsic principled unity and cannot have a unique character that is specified by the properties that have a principled connection to the kind and are understood to be aspects of being that kind of thing. The lexical/phrasal distinction is important for the theory of conceptual form because it captures the distinction between intrinsic principled unities and nonprincipled unities, and this distinction has many consequences for how we think and talk about things.

From the perspective of conceptual form, it would be convenient if the syntax provides the means for combining lexically expressible concepts such as BIRD, BLACKBIRD, and BLACK, which involve intrinsic principled unities into complex concepts such as BLACK BIRD, which do not. If, however, as many modern approaches to generative syntax assume, the terminals of syntactic structures are roots that are not wordlike representations, there would have to be some other way of reconstructing the distinction between the linguistic elements that are understood to provide the means for thinking and talking about intrinsic principled unities and those elements that provide the means for thinking and talking about things that do not possess an intrinsic principled unity.

6 New Perspectives on Generics

1. See Prasada (2010) for a more detailed discussion of how the interpretation of the indefinite singular generics that characterize kinds differs from that of indefinite singular generics that characterize ad hoc subkinds.

2. A large body of work by Daniel Kahneman and Amos Tversky has documented reasoning errors and fallacies that appear to arise from the use of a so-called representativeness heuristic, according to which we assume that every instance or sample of a category will have the properties that are characteristic of the category (e.g., Kahneman, Slovic, & Tversky, 1982). It is worth noting that given the constraints on the use of the definite singular generic, this is a perfectly rational way to reason. Furthermore, our conceptual systems are sensitive to the constraints on this mode of generalization, as evidenced by the constraints on the use of the definite generic in English. This suggests that things go awry when participants are asked to solve problems for which they do not have the relevant statistical tools and thus are forced to use mechanisms that they have available for solving similar problems in well-specified conditions outside those conditions. If this is correct, our cognitive architecture does not contain a representativeness *heuristic*, but it does contain a representative mode of generalization with implicitly known limits on its applicability.

7 Explaining Systematic Polysemy

1. The sketch of the conceptual form of kind concepts in Ritchie and Prasada (2024) abstracts away from those characteristics of conceptual form that are not directly relevant to the account of systematic polysemy, and thus it does not include some of the components of conceptual form included in the discussion here and in chapters 4 and 5 as those are needed to explain many other conceptual and linguistic phenomena.

2. I am assuming that OPM is represented as a lexically expressible kind concept rather than as the phrasal ad hoc kind concept ORGANIZATION THAT EXISTS FOR THE SAKE OF PRODUCING A MAGAZINE, in which case the two kind representations are obviously formally related, with MAGAZINE being a constituent of the ad hoc kind concept. One reason to think that OPM is a lexically expressible kind representation like DOG or MAGAZINE is that it appears to involve a principled unity, and thus the kind has principled connections to various properties such as having an editor and having writers. Phrasal (ad hoc) kind concepts cannot be assumed to represent a principled unity, and thus they do not have properties with a principled connection to them (see chapter 5). Furthermore, the fact that OPMs can be referred to via the lexical item "magazine" shows that it is, in fact, a lexically expressible kind concept. An important question for future research is to investigate whether the conceptual form of kind concepts such as OPM differs from both the conceptual form of phrasal concepts such as BLACK BIRD and simple lexically expressible concepts like BIRD, and, if so, how the instructions provided by the conceptual form of compound kind concepts such as OPM or BLACKBIRD differ from those provided by the conceptual form of phrasal and simple kind concepts.

8 Metaphorical Perspectives

1. Glucksberg (2011) proposes how these phenomena can potentially be explained by the categorization approach to metaphor via the mechanism of dual reference. As discussed in this chapter, the dual-reference mechanism envisioned by the categorization approach to metaphor may be unworkable.

9 Acquiring Perspectives

1. For discussions of some differences concerning the scope and content of core knowledge systems, see Revencu and Csibra (2023), Hamlin (2023), and Spelke (2023).

2. One reason to prefer this interpretation of the infant data is that it avoids the problems engendered by the idea that infants are forming superordinate sortal-kind concepts, including the following: (1) If infants are really forming superordinate sortal-kind concepts, what prevents them from learning superordinate terms for

those categories? (2) There is a hierarchical kind-of relation between basic-level and superordinate categories. Furthermore, every instance of the superordinate category must be an instance of some basic-level category (see chapter 5). Do we really want to attribute all that knowledge to infants in these experiments? If not, it raises the questions of in what sense are the infants representing a superordinate sortal-kind, as well as how children eventually acquire knowledge of the hierarchical relation between basic-level and superordinate sortal-kind concepts.

10 The Nature of Conceptual Knowledge

1. Laurence and Margolis (2012) argue convincingly that despite what proponents of such views may hold, such accounts presuppose the existence of at least some general concepts.

2. This makes it clear that not only do we not need to form a conceptual representation of particulars prior to forming a kind representation, we need not have a perceptual representation of a particular prior to forming a kind representation that would apply to perceivable particulars.

3. Some exceptions include Laurence and Margolis (2012), Carey (2009), and Strevens (2012). All these studies differ from the theory of conceptual form in many ways, including the fact that, unlike the theory of conceptual form, they view conceptual knowledge as always being revisable.

4. A third perspective on domain-specificity in the brain is that it depends on patterns of connectivity between brain regions (Mahon & Caramazza, 2011; Mahon, 2023). I will discuss that view and its relation to conceptual form in chapter 11, as it has no counterpart in existing theories of conceptual representation.

11 New Horizons for the Study of Conceptual Form

1. Although the details of exactly how their model does this are important and interesting, I refer the reader to the paper itself for them, as those details are not directly relevant to the points that I want to make here.

References

Adger, D. (2019). *Language unlimited: The science behind our most creative power.* Oxford University Press.

Ahn, W. (1999). Effect of Causal Structure on Category Construction. *Memory & Cognition, 27,* 1008–1023.

Akhlaghpour, H. (2022). An RNA-based theory of natural universal computation. *Journal of Theoretical Biology, 537,* 1–19. https://doi.org/10.1016/j.jtbi.2021.110984

Apresjan, J. D. (1974). Regular polysemy. *Linguistics, 12*(142), 5–32.

Asher, N. (2011). *Lexical meaning in context: A web of words.* Cambridge University Press.

Asher, N., & Pelletier, F. J. (2012). More truths about generic truth. In A. Mari, C. Beyssade, & F. Del Prete (Eds.), *Genericity* (pp. 313–333). Oxford University Press.

Aslanov, I., & Guerra, E. (2023). Tautological formal explanations: Does prior knowledge affect their satisfiability? *Frontiers in Psychology, 14,* 1258985.

Astuti, R., Solomon, G. E., & Carey, S. (2004). Constraints on conceptual development: A case study of the acquisition of folkbiological and folksociological knowledge in Madagascar. *Monographs of the Society for Research in Child Development,* 1–161.

Ayers, M. (1997). Is *physical object* a sortal concept? A reply to Xu. *Mind & Language, 12*(3–4), 393–405. https://doi.org/10.1111/1468-0017.00053

Baker, M. (2003). *Lexical categories: Verbs, nouns, and adjectives.* Cambridge University Press.

Baker, M. C. (2008). *The atoms of language: The mind's hidden rules of grammar.* Basic Books.

Barsalou, L. W. (1985). Ideals, central tendency, and frequency of instantiation as determinants of graded structure in categories. *Journal of Experimental Psychology: Learning, Memory, and Cognition, 11*(4), 629–654.

Barsalou, L. W. (2008). Grounded cognition. *Annual Review of Psychology, 59*, 617–645. https://doi.org/10.1146/annurev.psych.59.103006.093639

Barsalou, L. W. (2012). The human conceptual system. In M. Spivey, K. McRae, & M. Joanisse (Eds.), *The Cambridge handbook of psycholinguistics* (pp. 239–258). Cambridge University Press.

Block, N. (1986). Advertisement for a semantics for psychology. *Midwest Studies in Philosophy, 10*, 615–678.

Blok, S. V., Newman, G. E., & Rips, L. J. (2007). Out of sorts? Some remedies for theories of object concepts: A reply to Rhemtulla and Xu (2007). *Psychological Review, 114*(4), 1096–1102. https://doi.org/10.1037/0033-295X.114.4.1096

Bowdle, B. F., & Gentner, D. (2005). The career of metaphor. *Psychological Review, 112*(1), 193–216. https://doi.org/10.1037/0033-295X.112.1.193

Brentano, F. (1982). *Deskriptive Psychologie* (Descriptive psychology). Edited by R. M. Chisholm and W. Baumgartner. Felix Meiner.

Bublitz, D., & Prasada, S. (2013). Normative dimensions of generic knowledge. Poster presented at the 35th Annual Cognitive Science Conference, July 31–August 3, Berlin.

Burnett, R. C., Medin, D. L., Ross, N. O., & Blok, S. V. (2005). Ideal is typical. *Canadian Journal of Experimental Psychology/Revue canadienne de psychologie expérimentale, 59*(1), 3–10. https://doi.org/10.1037/h0087453

Camp, E. (2019). Imaginative frames for scientific inquiry: Metaphors, telling facts, and just-so stories. In P. Godfrey-Smith and A. Levy (Eds.), *The scientific imagination* (pp. 304–336). Oxford University Press.

Campbell, J., & Hall, D. G. (2022). The scope of infants' early object word extensions. *Cognition, 228*, 105210.

Capitani, E., Laiacona, M., Mahon, B., & Caramazza, A. (2003). What are the facts of category-specific deficits? A critical review of the clinical evidence. *Cognitive Neuropsychology, 20*, 213–261.

Caramazza, A., & Shelton, J. R. (1998). Domain-specific knowledge systems in the brain: The animate–inanimate distinction. *Journal of Cognitive Neuroscience, 10*(1), 1–34. https://doi.org/10.1162/089892998563752

Carey, S. (1985). *Conceptual change in childhood.* MIT Press.

Carey, S. (2009). *The origin of concepts.* Oxford University Press. https://doi.org/10.1093/acprof:oso/9780195367638.001.0001

Carey, S. (2015). Why theories of concepts should not ignore the problem of acquisition. In E. Margolis & S. Laurence (Eds.), *The conceptual mind: New directions in the study of concepts* (pp. 415–454). MIT Press.

Carey, S., & Spelke, E. (1994). Domain-specific knowledge and conceptual change. In L. A. Hirschfeld & S. A. Gelman (Eds.), *Mapping the mind: Domain specificity in cognition and culture* (pp. 169–200). Cambridge University Press. https://doi.org/10.1017/CBO9780511752902.008

Carey, S., & Xu, F. (1999). Kinds and sortals: An appreciation of John Macnamara. In R. S. Jackendoff, P. Bloom, & K. Wynn (Eds.), *Language, logic, and concepts: Essays in memory of John Macnamara* (pp. 311–335). MIT Press.

Carey, S. & Xu, F. (2001). Infants' knowledge of objects: Beyond object files and object tracking. *Cognition, 80*, 179–213.

Carlson, G. (1977). *Reference to kinds in English.* PhD thesis, University of Massachusetts at Amherst.

Carlson, G. N. (1995). Truth conditions of generic sentences: Two contrasting views. In G. N. Carlson & F. J. Pelletier (Eds.), *The generic book* (pp. 224–237). University of Chicago Press.

Carlson, G. (2010). Generics and concepts. In F. J. Pelletier (Ed.), *Kinds, things & stuff* (pp. 16–35) Oxford University Press.

Carlson, G. N., & Pelletier, F. J. (1995). *The generic book.* University of Chicago Press.

Chierchia, G. (1998). Reference to kinds across language. *Natural Language Semantics, 6*, 339–405.

Chomsky, N. (1986). *Knowledge of language: Its Nature, Origin, and Use.* New York: Praeger.

Chomsky, N. (2000). *New horizons in the study of language and mind.* Cambridge University Press.

Chomsky, N. (2003). Reply to Rey. In L. M. Antony & N. Hornstein (Eds.), *Chomsky and his critics* (pp. 274–287). Blackwell.

Chomsky, N. (2008). On phases. In R. Frieden, C. P. Otero, & M. L. Zubizarreta (Eds.), *Foundational issues in linguistic theory: Essays in honor of Jean-Roger Vergnault* (pp. 133–166). MIT Press.

Chomsky, N. (2009). Conclusion. In M. Piattelli-Palmarini, J. Uriagereka, & P. Salaburu (Eds.), *Of minds and language: A dialogue with Noam Chomsky in the Basque Country* (pp. 379–409). Oxford University Press.

Chomsky, N. (2023). Genuine explanation and the strong minimalist thesis. *Cognitive Semantics, 8*, 347–365.

Chomsky, N., Seely, T. D., Berwick, R. C., Fong, S., Huybregts, M. A. C., Kitahara, H., . . . , & Sugimoto, Y. (2023). *Merge and the strong minimalist thesis.* Cambridge University Press.

Cimpian, A. (2016). The privileged status of category representations in early development. *Child Development Perspectives, 10*(2), 99–104. https://doi.org/10.1111/cdep.12166

Cohen, A. (1999). Generics, frequency adverbs, and probability. *Linguistics and Philosophy, 22*, 221–253.

Cohen, A. (2001). On the generic use of indefinite singulars. *Journal of Semantics, 18*, 183–209.

Cohen, A. (2002). Genericity. *Linguistische Berichte Sonderheft, 10*, 59–89.

Collins, J. (2015). Generics as a unitary psychological phenomenon: An argument from linguistic diversity. *Ratio, 28*(4), 369–394.

Collins, J. (2017). The copredication argument. *Inquiry, 60*(7), 675–702.

Colunga, E., & Smith, L. B. (2005). From the lexicon to expectations about kinds: A role for associative learning. *Psychological Review, 112*(2), 347–382. https://doi.org/10.1037/0033-295X.112.2.347

Cree, G. S., & McRae, K. (2003). Analyzing the factors underlying the structure and computation of the meaning of *chipmunk, cherry, chisel, cheese,* and *cello* (and many other such concrete nouns). *Journal of Experimental Psychology: General, 132*, 163–201. https://doi.org/10.1037/0096-3445.132.2.163

Croteau, J., Cheries, E., & Xu, F. (2024). The development of kind concepts: Insights from object individuation. *Psychological Review*. https://doi.org/10.1037/rev0000527

Dahl, O. (1975). On generics. In E. L. Keenan (Ed.), *Formal semantics of natural language* (pp. 99–111). Cambridge University Press.

Dayal, V. (2004). Number marking and (in)definiteness in kind terms. *Linguistics and Philosophy, 27*, 393–450.

Declerck, R. (1991). The origins of genericity. *Linguistics, 29*, 79–102.

Devlin, J. T., Gonnerman, L. M., Andersen, E. S., & Seidenberg, M. S. (1998). Category-specific semantic deficits in focal and widespread brain damage: A computational account. *Journal of Cognitive Neuroscience, 10*(1), 77–94.

Dölling, J. (1995). Ontological domains, semantic sorts and systematic ambiguity. *International Journal of Human-Computer Studies, 43*, 785–807.

Embick, D., & Poeppel, D. (2005). Mapping syntax using imaging: Prospects and problems for the study of neurolinguistic computation. In K. Brown (Ed.), *Encyclopedia of language and linguistics* (2nd ed., pp. 484–486). Elsevier.

Feigenson, L., & Halberda, J. (2008). Conceptual knowledge increases infants' memory. *Proceedings of the National Academy of Sciences, 105*(29), 9926–9930.

Fenton, T., & Prasada, S. (2024). Some evidence for the formal characteristics of kind and instance-of-kind representations. Poster presented at the 50th Meeting of the Society of Philosophy and Psychology, June 20–23, West Lafayette, IN.

Fenton, T., & Prasada, S. (2025). What systematic polysemy reveals about hierarchical structure in semantic memory. Poster presented at the 51st Meeting of the Society of Philosophy & Psychology, June 19–21, Ithaca, NY.

Ferenz, K. S. (2000). *The role of nongeometric information in spatial language.* Doctoral dissertation, Dartmouth College.

Fodor, Jerry A. (1975). *The language of thought.* Harvard University Press.

Fodor, J. (1998a). *Concepts: Where cognitive science went wrong.* Oxford University Press.

Fodor, J. (1998b). When is a dog a DOG? Review of *Similarity and symbols in human thinking* edited by Steven A. Sloman and Lance J. Rips. *Nature, 396*(6709), 325.

Fodor, J. A., & Lepore, E. (1998). The emptiness of the lexicon: Reflections on James Pustejovsky's *The generative lexicon. Linguistic Inquiry, 29*(2), 269–288.

Fodor, J. A., & Pylyshyn, Z. W. (1988). Connectionism and cognitive architecture: A critical analysis. *Cognition, 28*(1–2), 3–71. https://doi.org/10.1016/0010-0277(88)90031-5

Foraker, S., & Murphy, G. L. (2012). Polysemy in sentence comprehension: Effects of meaning dominance. *Journal of Memory and Language, 67*(4): 407–425. https://doi.org/10.1016/j.jml.2012.07.010

Foster-Hanson, E., & Lombrozo, T. (2022). How "is" shapes "ought" for folk-biological concepts. *Cognitive Psychology, 139,* 1–28. https://doi.org/10.1016/j.cogpsych.2022.101507

Foster-Hanson, E., & Rhodes, M. (2019). Is the most representative skunk the average or the stinkiest? Developmental changes in representations of biological categories. *Cognitive Psychology, 110,* 1–15. https://doi.org/10.1016/j.cogpsych.2018.12.004

Frisson, S. (2015). About bound and scary books: The processing of book polysemies. *Lingua, 157,* 17–35.

Gallistel, C. R. (2017). The coding question. *Trends in Cognitive Sciences, 21,* 498–508. https://doi.org/10.1016/j.tics.2017.04.012

Gallistel, C. R. (2021). The physical basis of memory. *Cognition, 213,* 1–6. https://doi.org/10.1016/j.cognition.2020.104533

Gelman, R. (1990). First principles organize attention to and learning about relevant data: Number and the animate-inanimate distinction as examples. *Cognitive Science, 14,* 79–106.

Gelman, S. A. (2003). *The essential child: Origins of essentialism in everyday thought.* Oxford University Press.

Gelman, S. A. (2004). Psychological essentialism in children. *Trends in Cognitive Sciences, 8*(9), 404–409.

Gelman, S. A., Cimpian, A., & Roberts, S. O. (2018). How deep do we dig? Formal explanations as placeholders for inherent explanations. *Cognitive Psychology, 106,* 43–59.

Gelman, S. A., Goetz, P. J., Sarnecka, B. W., & Flukes, J. (2008). Generic language in parent- child conversations. *Language Learning and Development, 4*(1), 1–31. https://doi.org/10.1080/15475440701542625

Gelman, S. A., & Raman, L. (2003). Preschool children use linguistic form class and pragmatic cues to interpret generics. *Child Development, 74*(1), 308–325. https://doi.org/10.1111/1467-8624.00537

Gelman, S. A., & Rhodes, M. (2012). Two-thousand years of stasis. In E. M. Evans, G. M. Sinatra, K. S. Rosengren, & S. K. Brem (Eds.), *Evolution challenges: Integrating research and practice in teaching and learning about evolution* (pp. 200–207). Oxford Univerisity Press.

Gelman, S. A., & Tardif, T. (1998). A cross-linguistic comparison of generic noun phrases in English and Mandarin. *Cognition, 66*(3), 215–248. https://doi.org/10.1016/S0010-0277(98)00021-3

Gentner, D., & Markman, A. B. (1994). Structural alignment in comparison: No difference without similarity. *Psychological Science, 5*(3), 152–158.

Gibbs, R. W., Jr. (Ed.). (2008). *Cambridge handbook of metaphor and thought.* Cambridge University Press. https:// doi.org/10.1017/CBO9780511816802

Gleitman, L. (1990). The structural sources of verb meanings. *Language Acquisition: A Journal of Developmental Linguistics, 1*(1), 3–55. https://doi.org/10.1207/s15327817la0101_2

Glucksberg, S. (2008). How metaphors create categories—quickly. In R. W. Gibbs, Jr. (Ed.), *The Cambridge handbook of metaphor and thought* (pp. 67–83). Cambridge University Press. https://doi.org/10.1017/CBO9780511816802.006

Glucksberg, S. (2011). Understanding metaphors: The paradox of unlike things compared. In Khurshid Ahmad (Ed.), *Affective computing and sentiment analysis: Emotion, metaphor and terminology* (pp. 1–12). Springer.

Glucksberg, S., & Haught, C. (2006). Can Florida become like the next Florida? When metaphoric comparisons fail. *Psychological Science, 17,* 935–938. https://doi.org/10.1111/j.1467-9280.2006.01807.x

Glucksberg, S., & Keysar, B. (1990). Understanding metaphorical comparisons: Beyond similarity. *Psychological Review, 97*(1), 3–18. https://doi.org/10.1037/0033-295X.97.1.3

Goldberg, R. F., & Thompson-Schill, S. L. (2009). Developmental "roots" in mature biological knowledge. *Psychological Science, 20*, 480–487.

Goldin-Meadow, S., Gelman, S. A., & Mylander, C. (2005). Expressing generic concepts with and without a language model. *Cognition, 96*(2), 109–126. https://doi.org/10.1016/j.cognition.2004.07.003

Goldstone, R. L., Kersten, A., & Carvalho, P. F. (2018). Categorization and concepts. In *Stevens' handbook of experimental psychology and cognitive neuroscience, vol. 3* (pp. 275–317). Wiley Online Library. https://doi.org/10.1002/9781119170174

Goodman, N. (1955). *Fact, fiction, & forecast*. Harvard University Press.

Gopnik, A., & Meltzoff, A. N. (1997). *Words, thoughts, and theories*. MIT Press.

Graham, S. A., Gelman, S. A., & Clarke, J. (2016). Generics license 30-month-olds' inferences about the atypical properties of novel kinds. *Developmental Psychology, 52*(9), 1353–1362. https://doi.org/10.1037/dev0000183

Green, E. J., & Quilty-Dunn, J. (2021). What is an object file? *British Journal for the Philosophy of Science, 72*, 665–699.

Grimshaw, J. (1981). Form, function, and the language acquisition device. In C. L. Baker & John J. McCarthy (Eds.), *The logical problem of language acquisition* (pp. 165–182). MIT Press.

Gupta, A. (1980). *The logic of common nouns: An investigation in quantified modal logic*. Yale University Press.

Hall, D. G., & Waxman, S. R. (1993). Assumptions about word meaning: Individual and basic- level kinds. *Child Development, 64*, 1550–1570.

Hamlin, J. K. (2023). Core morality? Or merely core agents and social beings? A response to Spelke's *What babies know*. *Mind & Language, 38*(5), 1323–1335.

Hampton, J. A. (1997). Psychological representations of concepts. In M. A. Conway (Ed.), *Cognitive models of memory* (pp. 81–110). Psychology Press.

Haught, C. (2013). A tale of two tropes: How metaphor and simile differ. *Metaphor and Symbol, 28*(4), 254–274.

Haward, Paul. 2020. *The developmental origins of the formal structure of kind representations*. Doctoral dissertation, Harvard University, Graduate School of Arts & Sciences.

Haward, P., Carey, S., & Prasada, S. (2021). The formal structure of kind representations. *Cognitive Science, 45*(10), e13040.

Haward, P., Wagner, L., Carey, S., & Prasada, S. (2018). The development of principled connections and kind representations. *Cognition, 176,* 255–268.

Hebart, M. N., Zheng, C. Y., Pereira, F., & Baker, C. I. (2020). Revealing the multidimensional mental representations of natural objects underlying human similarity judgements. *Nature Human Behaviour, 4*(11), 1173–1185. https://doi.org/10.1038/s41562-020-00951-3

Holyoak, K. J., & Stamenković, D. (2018). Metaphor comprehension: A critical review of theories and evidence. *Psychological Bulletin, 144*(6), 641–671. https://doi.org/10.1037/bul0000145

Hu, Z., van Paridon, J., & Lupyan, G. (2024). Failures and successes in learning a core conceptual distinction from the statistics of language. In J. Nölle, L. Raviv, K. E. Graham, S. Hartmann, Y. Jadoul, M. Josserand, T. Matzinger, et al. (Eds.), *The evolution of language: Proceedings of the 15th International Conference (EVOLANG XV).* https://doi.org/10.17617/2.3587960.

Hull, David L. (1964). The effect of essentialism on taxonomy—two thousand years of stasis. *British Journal for the Philosophy of Science, 15,* 314–326.

Hutchison, K. (1991). Dormitive virtues, scholastic qualities, and the new philosophies. *History of Science, 29*(3), 245–278.

Imai, M., & Gentner D. (1997). A cross-linguistic study of early word meaning: Universal ontology and linguistic influence. *Cognition, 62,* 169–200. https://doi.org/10.1016/s0010-0277(96)00784-6

Isaac, E. (2015). What makes a Picasso, a Picasso? Producer–product polysemy and its implications for theories of lexical representation. MA thesis, Hunter College, City University of New York (CUNY).

Jackendoff, R. (1983). *Semantics and cognition.* MIT Press.

Jackendoff, R. (1987). *Consciousness and the computational mind.* MIT Press.

Jackendoff, R. (2002). *Foundations of language: How language connects to the brain, the world, evolution, and thinking.* Oxford University Press.

Jones, L. L., & Estes, Z. (2006). Roosters, robins, and alarm clocks: Aptness and conventionality in metaphor comprehension. *Journal of Memory and Language, 55*(1), 18–32. https://doi.org/10.1016/j.jml.2006.02.004

Kahneman, D., Slovic, P., & Tversky, A. (1982). *Judgment under uncertainty: Heuristics and biases.* Cambridge University Press.

Kahneman, D., Treisman, A., & Gibbs, B. J. (1992). The reviewing of object files: Object-specific integration of information. *Cognitive Psychology, 24*(2), 175–219. https://doi.org/10.1016/0010-0285(92)90007-O

Kazanina, N., & Poeppel, D. (2023). The neural ingredients for a language of thought are available. *Trends in Cognitive Sciences, 11*(27), 996–1007. https://doi.org/10.1016/j.tics.2023.07.012

Keil, F. C. (1989). *Concepts, kinds, and cognitive development.* Bradford.

Keil, F. C. (1994). The birth and nurturance of concepts by domains: The origins of concepts of living things. In L. A. Hirschfeld & S. A. Gelman (Eds.), *Mapping the mind: Domain specificity in cognition and culture* (pp. 234–254). Cambridge University Press. https://doi.org/10.1017/CBO9780511752902.010

Keil, F. C. (2003). Folkscience: Coarse interpretations of a complex reality. *Trends in Cognitive Sciences, 7*(8), 368–373. https://doi.org/10.1016/s1364-6613(03)00158-x

Keil, F. C., Smith, W. C., Simons, D. J., & Levin, D. T. (1998). Two dogmas of conceptual empiricism: Implications for hybrid models of the structure of knowledge. *Cognition, 65*(2–3), 103–135. https://doi.org/10.1016/S0010-0277(97)00041-3

Kelemen, D. (1999). The scope of teleological thinking in preschool children. *Cognition, 70*(3), 241–272. https://doi.org/10.1016/S0010-0277(99)00010-4

Kelemen, D., Rottman, J., & Seston, R. (2013). Professional physical scientists display tenacious teleological tendencies: Purpose-based reasoning as a cognitive default. *Journal of Experimental Psychology: General, 142*, 1074–1083.

Kibbe, M.M., & Leslie, A.M. (2011). What do infants remember when they forget? Location and identity in 6-month-olds' memory for objects. *Psychological Science 22*(12), 1500-1505.

Kibbe, M. M., & Leslie, A. M. (2019). Conceptually rich, perceptually sparse: Object representations in 6-month-old infants' working memory. *Psychological Science, 30*, 362–375. https://doi.org/10.1177/0956797618817754

Kiefer, M., & Pulvermüller, F. (2012). Conceptual representations in mind and brain: Theoretical developments, current evidence, and future directions. *Cortex, 48*, 805–825. https://doi.org/10.1016/j.cortex.2011.04.006

Klein, D. E., & Murphy, G. L. (2001). The representation of polysemous words. *Journal of Memory and Language, 45*(2), 259–282.

Klepousniotou, E. (2002). The processing of lexical ambiguity: Homonymy and polysemy in the mental lexicon. *Brain and Language, 81*(1–3), 205–223. https://doi.org/10.1006/brln.2001.2518

Knobe, J. (under review). *In a deeper sense.* Yale University.

Knobe, J., Prasada, S., & Newman, G. E. (2013). Dual character concepts and the normative dimension of common sense conception. *Cognition, 127*, 242–257.

Knowlton, T., Hunter, T., Odic, D., Wellwood, A., Halberda, J., Pietroski, P., & Lidz, J. (2021). Linguistic meanings as cognitive instructions. *Annals of the New York Academy of Sciences, 1500*(1), 134–144.

Krifka, M., Pelletier, F. J., Carlson, G., ter Meulen, A., Chierchia, G., & Link, G. (1995). Genericity: An introduction. In Greg N. Carlson & Francis Jeffry Pelletier (Eds.), *The generic book* (pp. 1–124). University of Chicago Press.

Lake, B. M., Salakhutdinov, R., & Tenenbaum, J. B. (2015). Human-level concept learning through probabilistic program induction. *Science, 350*(6266), 1332–1338.

Lakoff, G., & Turner, M. (1989). *More than cool reason: A field guide to poetic metaphor.* University of Chicago Press. https://doi.org/10.7208/chicago/9780226470986.001.0001

LaTourette, A., Chan, D. M., & Waxman, S. 2023. A principled link between object naming and representation is available to infants by seven months of age. *Scientific Reports, 13,* 14328.

Laurence, S., & Margolis, E. (2012). Abstraction and the origin of general ideas. *Philosophers' Imprint, 12,* 1–22.

Laurence, S., & Margolis, E. (2024). *The building blocks of thought: A rationalist account of the origins of concepts.* Oxford University Press.

Lawler, J. (1973). Studies in English generics. *University of Michigan Papers in Linguistics, 1*(1).

Lazaridou-Chatzigoga, D., Katsos, N., & Stockall, L. (2019). Generalizing about striking properties: Do glippets love to play with fire? *Frontiers in Psychology, 10* (August 28), Article 1971. https://www.frontiersin.org/articles/10.3389/fpsyg.2019.01971

Lee, S. H., Ji, Y., & Papafragou, A. (2024). Signatures of individuation across objects and events. *Journal of Experimental Psychology: General, 153*(8), 1997–2012.

Legare, C. H., & Shtulman, A. (2018). Explanatory pluralism across cultures and development. In J. Proust & M. Fortier (Eds.), *Interdisciplinary approaches to metacognitive diversity* (pp. 415–432). Oxford University Press.

Leonard, N., & Rips, L. J. (2015). Identifying and counting objects: The role of sortal concepts. *Cognition, 145,* 89–103.

Leslie, A. M. (1987). Pretense and representation: The origins of "theory of mind." *Psychological Review, 94*(4), 412–426. https://doi.org/10.1037/0033-295X.94.4.412

Leslie, A. M., Xu, F., Tremoulet, P. D., & Scholl, B. J. (1998). Indexing and the object concept: Developing "what" and "where" systems. *Trends in Cognitive Sciences, 1.* https://doi.org/10.1016/s1364-6613(97)01113-3

Leslie, S.-J. (2007). Generics and the structure of the mind. *Philosophical Perspectives, 21*(1), 375–403. https://doi.org/10.1111/j.1520-8583.2007.00138.x

Leslie, S.-J. (2008). Generics: Cognition and acquisition. *Philosophical Review, 117*(1), 1–47. https://doi.org/10.1215/00318108-2007-023

Leslie, S.-J., Khemlani, S., Prasada, S., & Glucksberg, S. (2009). Conceptual and linguistic distinctions between singular and plural generics. In N. Taatgen, H. van Rijn, J. Nerbonne, & L. Schomaker (Eds.), *Proceedings of the 31st Annual Conference of the Cognitive Science Society*. (pp. 479–484). Cognitive Science Society.

Leslie, S.-J., & Lerner, A. (2022). Generic generalizations. In *Stanford encyclopedia of philosophy*. https://plato.stanford.edu/entries/generics/

Leus, O. A. (2021). Representation of sex-linked properties: Implications for conceptual representation and generics. *CUNY Academic Works*. https://academicworks.cuny.edu/hc_sas_etds/708

Li, P., & Abarbanell, L. (2019). Alternative spin on phylogenetically inherited spatial reference frames. *Cognition, 191*, 103983.

Li, P., Dunham, Y., & Carey S. (2009). Of substance: The nature of language effects on entity construal. *Cognitive Psychology, 58*(4), 487–524.

Liu, Q., & Lupyan, G. (2023). Cross-domain semantic alignment: Concrete concepts are more abstract than you think. *Philosophical Transactions of the Royal Society B, 378*(1870), 20210372.

Löhr, G., & Michel, C. (2022). Copredication in context: A predictive processing approach. *Cognitive Science, 46*, 1–27.

Lombrozo, T. (2009). Explanation and categorization: How "why?" informs "what?" *Cognition, 110*(2), 248–253. https://doi.org/10.1016/j.cognition.2008.10.007

Lombrozo, T., & Carey, S. (2006). Functional explanation and the function of explanation. *Cognition, 99*(2), 167–204. https://doi.org/10.1016/j.cognition.2004.12.009

Luck, S. J., & Vogel, E. K. (1997). The capacity of visual working memory for features and conjunctions. *Nature, 390*(6657), 279–281.

Lynch, E. B., Coley, J. D., & Medin, D. L. (2000). Tall is typical: Central tendency, ideal dimensions, and graded category structure among tree experts and novices. *Memory & Cognition, 28*(1), 41–50. https://doi.org/10.3758/BF03211575

MacGregor, L., Bouwsema, J., & Klepousniotou, E. (2015). Sustained meaning activation for polysemous but not homonymous words: Evidence from EEG. *Neuropsychologia, 68*, 126–138.

Macnamara, J. (1982). *Names for things: A study of human learning*. MIT Press.

Macnamara, J. (1986). *A border dispute: The place of logic in psychology*. MIT Press.

Macnamara, J. (1994). Logic and cognition. In J. Macnamara & G. E. Reyes (Eds.), *The logical foundations of cognition* (pp. 11–34). Oxford University Press.

Macnamara, J. (1999). *Through the rearview mirror: Historical reflections on psychology*. MIT Press.

Mahon, B. Z. (2023). Domain-specific connectivity drives the organization of object knowledge in the brain. *Handbook of Clinical Neurology, 187*, 221–244.

Mahon, B. Z., & Caramazza, A. (2009). Concepts & categories: A cognitive neuropsychological perspective. *Annual Review of Psychology, 60*, 27–51.

Mahon, B. Z., & Caramazza, A. (2011). What drives the organization of object knowledge in the brain? *Trends in Cognitive Sciences, 15*, 97–103.

Margolis, E. (1998). How to acquire a concept. *Mind & Language, 13*, 347–369.

Margolis, E., & Laurence, S. (1999). *Concepts: Core readings*. MIT Press.

Margolis, E., & Laurence, S. (2015). *The conceptual mind: New directions in the study of concepts*. MIT Press.

Margolis, E., & Laurence, S. (2023). Concepts. In E. N. Zalta & U. Nodelman (Eds.), *The Stanford encyclopedia of philosophy* (Fall 2023 ed.). Stanford University Press. https://plato.stanford.edu/archives/fall2023/entries/concepts/

Mari, A., Beyssade, C., & Del Prete, F. (2012). *Genericity*. Oxford University Press.

Martin, A. (2007). The representation of object concepts in the brain. *Annual Review Psychology, 58*, 25–45. https://doi.org/10.1146/annurev.psych.57.102904.190143

McPherson, L. M. P. (1995). *Identifying verbs early in language learning: The roles of action and argument structure*. PhD dissertation, McGill University, Montreal.

McRae, K., de Sa, V. R., & Seidenberg, M. S. (1997). On the nature and scope of featural representations of word meaning. *Journal of Experimental Psychology: General, 126*(2), 99–130. doi: 10.1037//0096-3445.126.2.99

Medin, D. L. (1989). Concepts and conceptual structure. *American Psychologist, 44*(12), 1469–81.

Medin, D. L., & Ortony, A. (1989). Psychological essentialism. In S. Vosniadou & A. Ortony (Eds.), *Similarity and analogical reasoning* (pp. 179–195). Cambridge University Press. https://doi.org/10.1017/CBO9780511529863.009

Medin, D. L., & Schaffer, M. M. (1978). Context theory of classification learning. *Psychological Review, 85*, 207–238. https://doi.org/10.1037/0033-295X.85.3.207

Mendia, J. A. (2019). Reference to *ad hoc* kinds. *Linguistics & Philosophy, 43*, 589–631.

Molière. (2012). *Le malade imaginere.* Translated by H. van Laun. Dover. Original work published in 1673.

Moravcsik, J. M. (1975). Aitia as a generative factor in Aristotle's philosophy. *Dialogue: Canadian Philosophical Review/Revue canadienne de philosophie, 14*(4), 622–638. https://doi.org/10.1017/S001221730002655X

Moravcsik, J. M. E. (1981). How do words get their meanings? *Journal of Philosophy, 78*, 5–24.

Moravcsik, J. M. E. (1998). *Meaning, creativity, and the partial inscrutability of the human mind.* CSLI Publications.

Muradoglu, M., Marchak, K. A., Gelman, S. A., & Cimpian, A. (2022). Formal explanations shape children's representations of animal kinds and social groups. *Developmental Psychology, 58*(12), 2322–2335. https://doi.org/10.1037/dev0001447

Murphy, E. (2019). Acceptability properties of abstract senses in copredication. In M. Bolognesi & G. Steen (Eds.), *Perspectives on abstract concepts: From cognitive processing to semantic representation and linguistic expression* (pp. 145–165). John Benjamins.

Murphy, E. (2021). Predicate order and coherence in copredication. *Inquiry, 67*(6), 1744–1780. https://doi.org/10.1080/0020174X.2021.1958054

Murphy, E. (2023). The Citadel itself: Defending semantic internalism. *Global Philosophy, 33*(1), 7.

Murphy, G. L. (1996). On metaphoric representation. *Cognition, 60*(2), 173–204. https://doi.org/10.1016/0010-0277(96)00711-1

Murphy, G. L. (2002). *The big book of concepts.* MIT Press.

Murphy, G. L. (2016). Is there an exemplar theory of concepts? *Psychonomic Bulletin & Review, 23*, 1035–1042.

Murphy, G. L., & Medin, D. L. (1985). The role of theories in conceptual coherence. *Psychological Review, 92*(3), 289–316. https://doi.org/10.1037/0033-295X.92.3.289

Naughtin, C. K., Mattingley, J. B., & Dux, P. E. (2016). Distributed and overlapping neural substrates for object individuation and identification in visual short-term memory. *Cerebral Cortex, 26*(2), 566–575.

Neufeld, E. (2022). Psychological essentialism and the structure of concepts. *Philosophy compass, 17*(5), e12823.

Noyes, A., Keil, F. C., Dunham, Y., & Ritchie, K. (2023). Same people, different group: Social structures are a central component of group concepts. *Cognition, 240*, 105567.

Nosofsky, R. M. (1988). Similarity, frequency, and category representations. *Journal of Experimental Psychology: Learning, Memory, and Cognition, 14*, 54–65. https://doi .org/10.1037/0278-7393.14.1.54

Ortega-Andrés, M., & Vicente, A. (2019). Polysemy and co-predication. *Glossa: A Journal of General Linguistics, 4*(1), Article 1. https://doi.org/10.5334/gjgl.564

Ortony, A. (1979). Beyond literal similarity. *Psychological Review, 86,* 161–180.

Pelletier, F. J., & Asher, N. (1997). Generics and defaults. In J. van Benthem & A. ter Meulen (Eds.), *Handbook of logic and language* (pp. 1125–1179). MIT Press.

Pietroski, P. M. (2018). *Conjoining meanings: Semantics without truth values.* Oxford University Press.

Pinker, S. (1984). *Language learnability and language development.* Harvard University Press.

Poeppel, D. (2012). The maps problem and the mapping problem: Two challenges for a cognitive neuroscience of speech and language. *Cognitive Neuropsychology, 29,* 34–55.

Poeppel, D., & Idsardi, W. (2022). We don't know how the brain stores anything, let alone words. *Trends in Cognitive Scences, 26,* 1054–1055. https://doi.org/10.1016/j.tics.2022.08.010

Prasada, S. (2005). Being near the ceramic, but not the mug: On the role of construal in spatial language. To appear in L. A. Carlson & E. van der Zee (Eds.), *Functional features in language and space: Insights from perception, categorization and development* (pp. 205–218). Oxford University Press.

Prasada, S. (2010). Conceptual representations and some forms of genericity. In F. J. Pelletier (Ed.), *Kinds, things & stuff* (pp. 36–59). Oxford University Press.

Prasada, S. (2012). Mechanisms for characterizing kinds and classes. *Recherches Linguistiques de Vincennes, 41,* 45–62.

Prasada, S. (2017). The scope of formal explanation. *Psychonomic Bulletin & Research, 24,* 1478–1487.

Prasada, S. (2021a). Formal explanation and mechanisms of conceptual representation. In L. Jansen & P. Sandstad (Eds.), *Neo-Aristotelian perspectives on formal causation* (pp. 269–286). Routledge.

Prasada, S. (2021b). The physical basis of conceptual representation—an addendum to. *Cognition, 214,* 104751.

Prasada, S., & Dillingham, E. M. (2006). Principled and statistical connections in common sense conception. *Cognition, 99*(1), 73–112. https://doi.org/10.1016/j.cognition.2005.01.003

Prasada, S., & Dillingham, E. M. (2009). Representation of principled connections: A window into the formal aspect of common sense conception. *Cognitive Science, 33*(3), 401–448. https://doi.org/10.1111/j.1551-6709.2009.01018.x

Prasada, S., Ferenz, K., & Haskell, T. (2002). Conceiving of entities as objects and stuff. *Cognition, 83*, 141–165.

Prasada, S., & Hall, D. G. (2019). Instance-of-object-kind representations. *Cognition, 189*, 209–220.

Prasada, S., Hennefield, L., & Otap, D. (2012). Conceptual and linguistic representations of kinds and classes. *Cognitive Science, 36*, 1224–1250.

Prasada, S., Khemlani, S., Leslie, S.-J., & Glucksberg, S. (2013). Conceptual distinctions amongst generics. *Cognition, 126*, 405–422.

Prasada, S., & Leus, O. A. (2024). *Ducks lay eggs. Lions have manes: The conceptual basis for minority characteristic generics*. Manuscript. Hunter College, City University of New York (CUNY).

Pustejovsky, J. (1995). *The generative lexicon*. MIT Press.

Pylyshyn, Z. (1989). The role of location indexes in spatial perception: A sketch of the FINST spatial-index model. *Cognition, 32*(1), 65–97. https://doi.org/10.1016/0010-0277(89)90014-0

Pylyshyn, Z. W. (2001). Visual indexes, preconceptual objects, and situated vision. *Cognition, 80*(1–2), 127–158. https://doi.org/10.1016/S0010-0277(00)00156-6

Quilty-Dunn, J. (2021). Polysemy and thought: Toward a generative theory of concepts. *Mind & Language, 36*(1), 158–185. https://doi.org/10.1111/mila.12328

Quilty-Dunn, J., Porot, N., & Mandelbaum, E. (2023). The best game in town: The reemergence of the language-of-thought hypothesis across the cognitive sciences. *Behavioral and Brain Sciences, 46*, e261. https://doi.org/10.1017/S0140525X22002849

Rabagliati, H., Marcus, G. F., & Pylkkanen, L. (2011). Rules, radical pragmatics and restrictions on regular polysemy. *Journal of Semantics, 28*(4), 485–512. https://doi.org/10.1093/jos/ffr005

Récanati, F. (2012). *Mental files*. Oxford University Press.

Rehder, B. (2003). A causal-model theory of conceptual representation and categorization. *Journal of Experimental Psychology: Learning, Memory, and Cognition, 29*, 1141–1159.

Rehder, B. (2017). Concepts as causal models: Classification. In M.R. Waldmann (Ed.), *The Oxford handbook of causal reasoning* (pp. 346–375). Oxford, UK: Oxford University Press.

Revencu, B., & Csibra, G. (2023). The missing link between core knowledge and language: Review of Elizabeth Spelke's *What babies know, volume 1* (2022). *Mind and Language, 38*(5), 1314–1322.

Rhemtulla, M., & Xu, F. (2007). Sortal concepts and causal continuity: Comment on Rips, Blok, and Newman (2006). *Psychological Review, 114*(4), 1087–1094. https://doi.org/10.1037/0033-295X.114.4.1087

Rips, L. (in press). *Taking shape: Mental concepts of physical objects*. Oxford University Press.

Rips, L. J., Blok, S., & Newman, G. (2006). Tracing the identity of objects. *Psychological Review, 113*(1), 1–30. https://doi.org/10.1037/0033-295X.113.1.1

Ritchie, K., & Knobe, J. (2020). Kindhood and essentialism: Evidence from language. In M. Rhodes (Ed.), *Advances in child development and behavior, 59*, 133–164. https://doi.org/10.1016/bs.acdb.2020.05.005

Ritchie, K., & Prasada, S. (2024). Explaining systematic polysemy: Kinds and individuation. *Inquiry*, 1–31.

Rivera, S., Prasad, S., & Prasada, S. (2023). Are formal explanations mere placeholders or pointers? *Cognition, 235*, 105407.

Roberts, S. O. (2022). Descriptive-to-prescriptive (D2P) reasoning: An early emerging bias to maintain the status quo. *European Review of Social Psychology, 33*(2), 289–322.

Rosch, E. (1975). Cognitive representations of semantic categories. *Journal of Experimental Psychology: General, 104*(3), 192–233. https://doi.org/10.1037/0096-3445.104.3.192

Rosch, E. (1978). Principles of categorization. In E. Rosch & B. Lloyd (Eds.), *Cognition and categorization* (pp. 27–48). Lawrence Erlbaum Associates.

Ryu, S. H., Yang, W., & Park, J. C. (2022). Flexible acceptance condition of generics from a probabilistic viewpoint: Towards formalization of the semantics of generics. *Journal of Psycholinguistic Research, 51*(6), 1209–1229.

Scholl, B. J. (2024). Perceptual (roots of) core knowledge [Commentary]. *Behavioral and Brain Sciences, 47*, e140. doi:10.1017/S0140525X23003023

Scholl, B. J., & Leslie, A. M. (1999). Modularity, development and "theory of mind." *Mind & Language, 14*(1), 131–153. https://doi.org/10.1111/1468-0017.00106

Shipley, E. F. (1993). Categories, hierarchies, and induction. *Psychology of learning and motivation*, 30, 265–301.

Shipley, E., & Shepperson, B. (1990). Countable entities: Developmental changes. *Cognition, 34*, 109–136. https://doi.org/10.1016/0010-0277(90)90041-H

Shtulman, A., & Valcarcel, J. (2012). Scientific knowledge suppresses but does not supplant earlier intuitions. *Cognition, 124*, 209–215.

Shtulman, A., & Young, A. G. (2024). Tempering the tension between science and intuition. *Cognition, 243*, 105680.

Sloman, S., Love, B., & Ahn, W. (1998). Feature centrality and conceptual coherence. *Cognitive Science, 22*, 189-228.

Sloman, S., Lombrozo, T., & Malt, B. (2007). Ontological commitments and domain specific categorisation. In M. J. Roberts (Ed.), *Integrating the mind* (pp. 105–129). Psychology Press.

Sobel, D. M., & Kirkham, N. Z. (2006). Blickets and babies: The development of causal reasoning in toddlers and infants. *Developmental Psychology, 42*(6), 1103–1115. https://doi.org/10.1037/0012-1649.42.6.1103

Soja N. N., Carey S., & Spelke E. S. (1991). Ontological categories guide young children's inductions of word meaning: Object terms and substance terms. *Cognition, 38*, 179–211. https://doi.org/10.1016/0010-0277(91)90051-5

Spaulding, S. (2022). Phenomenology of social explanation. *Phenomenology and the Cognitive Sciences*, 1–17. https://doi.org/10.1007/s11097-022-09854-2

Spelke, E. S. (1990). Principles of object perception. *Cognitive Science, 14*(1), 29–56. https://doi.org/10.1207/s15516709cog1401_3

Spelke, E. S. (2022). *What babies know: Core knowledge and composition,* Vol. 1. Oxford University Press.

Spelke, E. S. (2023). Core knowledge, language learning, and the origins of morality and pedagogy: Reply to reviews of *What babies know*. *Mind & Language, 38*(5), 1336–1350.

Spelke, E. S., & Kinzler, K. D. (2007). Core knowledge. *Developmental Science, 10,* 89–96.

Springer, K., & Keil, F. C. (1991). Early differentiation of causal mechanisms appropriate to biological and nonbiological kinds. *Child Development, 62*(4), 767–781.

Srinivasan, M., Chestnut, E., Li, P., & Barner, D. (2013). Sortal concepts and pragmatic inference in children's early quantification of objects. *Cognitive Psychology, 66,* 302–326. doi: 10.1016/j.cogpsych.2013.01.003

Srinivasan, M., & Rabagliati, H. (2015). How concepts and conventions structure the lexicon: Cross-linguistic evidence from polysemy. *Lingua, 157*, 124–152. https://doi.org/10.1016/j.lingua.2014.12.004

Srinivasan, M., & Snedeker, J. (2011). Judging a *book* by its cover and its contents: The representation of polysemous and homophonous meanings in four-year-old children. *Cognitive Psychology, 62*(4), 245–272. https://doi.org/10.1016/j.cogpsych.2011.03.002

Srinivasan, M., & Snedeker, J. (2014). Polysemy and the taxonomic constraint: Children's representation of words that label multiple kinds. *Language Learning and Development, 10*(2), 97–128.

Sterken, R. K. (2017). The meaning of generics. *Philosophy Compass, 12*, e12431.

Strevens, M. (2000). The essentialist aspect of naive theories. *Cognition, 74*(2), 149–175. https://doi.org/10.1016/S0010-0277(99)00071-2

Strevens, M. (2012). Theoretical terms without analytic truths. *Philosophical Studies, 160*, 167–190.

Studtmann, P. (2012). Aristotle's categorial scheme. In C. Shields (Ed.), *Oxford handbook of Aristotle* (pp. 63–80). Oxford University Press.

Syrett, K., & Aravind, A. (2021). Context sensitivity and the semantics of count nouns in the evaluation of partial objects by children and adults. *Journal of Child Language, 49*, 239–265. https://doi.org/10.1017/S0305000921000027

Tessler, M. H., & Goodman, N. D. (2019). The language of generalization. *Psychological Review, 126*(3), 395–436. https://doi.org/10.1037/rev0000142

Tworek, C. M., & Cimpian, A. (2016). Why do people tend to infer "ought" from "is"? The role of biases in explanation. *Psychological Science, 27*(8), 1109–1122. https://doi.org/10.1177/0956797616650875

Tyler, L. K., & Moss, H. E. (2001). Towards a distributed account of conceptual knowledge. *Trends in Cognitive Sciences, 5*(6), 244–252. https://doi.org/10.1016/S1364-6613(00)01651-X

Tyler, L. K., Moss, H. E., Durrant-Peatfield, M. R., & Levy, J. P. (2000). Conceptual structure and the structure of concepts: A distributed account of category-specific deficits. *Brain and Language, 75*(2), 195–231.

Umbach, C., & Gust, H. (2014). Similarity demonstratives. *Lingua, 149*, 74–93.

VanMarle, K., & Scholl, B. (2003). Attentive tracking of objects versus substances. *Psychological Science, 14*, 498–504.

Vasilyeva, N., & Lombrozo, T. (2020). Structural thinking about social categories: Evidence from formal explanations, generics, and generalization. *Cognition, 204*, 104383. https://doi.org/10.1016/j.cognition.2020.104383

Vendler, Z. (1971). Singular terms. In D. Steinberg and L. Jakovitz (Eds.), *Semantics: An interdisciplinary reader* (pp. 115–133). Cambridge University Press.

Warrington, E. K. (1981). Neuropsychological studies of verbal semantic systems. *Philosophical Transactions of the Royal Society of London. B, Biological Sciences, 295*(1077), 411–423.

Warrington, E. K., & McCarthy, R. (1983). Category specific access dysphasia. *Brain, 106*, 859–78. https://doi.org/10.1093/brain/106.4.859

Warrington, E. K., & McCarthy, R. A. (1987). Categories of knowledge: Further fractionations and an attempted integration. *Brain, 110*, 1273–1296. https://doi.org/10.1093/brain/110.5.1273

Warrington, E. K., & Shallice, T. (1984). Category specific semantic impairments. *Brain, 107*, 829–854. https://doi.org/10.1093/brain/107.3.829

Waxman, S. R., & Markow, D. B. (1995). Words as invitations to form categories: Evidence from 12-month-old infants. *Cognitive Psychology, 29*, 257–302.

Wellman, H. M., & Gelman, S. A. (1992). Cognitive development: Foundational theories of core domains. *Annual Review of Psychology, 43*, 337–375. https://doi.org/10.1146/annurev.ps.43.020192.002005

Wellwood, A., Hespos, S. J., & Rips, L. (2018). How similar are objects and events? *Acta Linguistica Academica, 65*(2–3), 473–501.

Wiggins, D. (1980). *Sameness and substance*. Harvard University Press.

Wiggins, D. (1997). Sortal concepts: A reply to Xu. *Mind and Language, 12*, 413–421.

Wiggins, D. (2001). *Sameness and substance renewed*. Cambridge University Press.

Xu, F. (1997). From Lot's wife to a pillar of salt: Evidence that physical object is a sortal concept. *Mind and Language, 12*, 365–392.

Xu, F. (2010). Count nouns, *sortal* concepts, and the nature of early words. In J. Pelletier (Ed.), *Kinds, things, and stuff* (pp 191–206). Oxford University Press.

Xu, Y., & Chun, M. (2006). Dissociable neural mechanisms supporting visual short-term memory for objects. *Nature, 440*, 91–95. https://doi.org/10.1038/nature04262

Yang, C. (2016). *The price of linguistic productivity: How children learn to break the rules of language*. MIT Press.

Ye, M. (2024). Individual concepts and personal identity judgement. MA thesis, CUNY Academic Works. https://academicworks.cuny.edu/hc_sas_etds/1125

Yu, X., & Lau, E. (2023). The binding problem 2.0: Beyond perceptual features. *Cognitive Science, 47*, e13244.

Zadeh, L. H. (1965). Fuzzy sets. *Information and control, 8*, 338–353.

Zhang, Y., Zhang, Y. Y., & Fang, F. (2020). Neural mechanisms of feature binding. *Science China Life Sciences, 63*, 926–928.

Zhu, R. (2021). Preschoolers' acquisition of producer-product metonymy. *Cognitive Development, 59*, Article 101075. https://doi.org/10.1016/j.cogdev.2021.101075

Zhu, S. D., Zhang, L. A., & von der Heydt, R. (2020). Searching for object indexicals in the visual cortex. *Journal of Neurophysiology, 123*(5), 1979–1994.

Index

Note: Figures and Tables are indicated by italicized page numbers.

Abstract categories
  ad hoc, 171–172
  metaphors and, 170–171
Abstract characteristics
  of instance-of-kind perspective, 38–42
  metaphors and, 170–171
Abstract distinguishing characteristics, 24
Abstract entities, 22
Abstract formal distinctions, 197
Abstract formal knowledge, 140–141
Abstract instances, 97, 156, 158
  kinds intrinsically containing, 89–93, *91*
Abstract principles, 205
Abstract terms, 93–94, *95*
Accidental generalizations, 139
Accidental properties, 185
Ad hoc abstract categories, 171–172
Ad hoc attributive categories, 170
Ad hoc categories, 170–172, 175–176
Ad hoc instances, 87, *88*, 89
  contextual information and, 104–105
  identifying, 105
  polysemy and, 162–163
Ad hoc kinds, 102–105
  polysemy and, 162–163
Ad hoc subkinds, 99–102, *101*, 105
  bare plural generics and, 135–136
  definite singular generics and, 137–138
  indefinite singular generics and, 136–137
Agent core knowledge system, 206
Amodal representations, 8–9
Animals, 247
  concepts in, 28
Animate kind concepts, 98, *99*, 252–253
Arbitrary instances, 140–142
Aristotle, 95, 139
  cognitive perspective on categories of, 261–263
  on explanation forms, 96, 273n8
Artifact kind concepts, 74, 95, 98, *99*, 253
Artists, names of, for works of, 161
Aspect-kind relations, 131
Aspect representations, 77, 114
Atomic primitives, 232–233
Attributions
  literal, 178–179
  metaphorical, 178
  principled connections and, 181
Axes, 231

Bare kind representations, 183
Bare plural generics, 135–136
Bare plural noun phrases, 140
Basic-level object kind concepts, 87, 89, 208
Bayesian program learning (BPL), 9, 248–250
Binary Encoder Representations from Transformers (BERT), 202
BPL. *See* Bayesian program learning
Brain damage, conceptual knowledge impairment from, 251
Brain mechanisms
  connectivity and computations in, 254
  distributed domain-specific hypothesis and, 252–254
  domain specificity of conceptual knowledge in, 251–255
  investigating conceptual representations in, 255–258
  underlying conceptual representation, 251–260

Canonical kind concepts, 104
Categories
  abstract, 170–171
  ad hoc, 171–172, 175–176
  of Aristotle, 261–263
  brain damage and, 251
  definitions approach, 6
  exemplar approach, 7–8
  ideal members of, 46
  pretending and, 175
  prototypes and representation of, 43, 46
  prototypes approach, 6–7
  psychological essentialism and, 11–12
  semantic bootstrapping and, 204
  as sets, 48–50
  summary descriptions and representing, 42–48
  topic, 181
  vehicle, 169–173, 176–177, 181
Categorization
  dual-reference mechanism and, 172–173
  explanation of properties and, 22, 40
  instance-of-kind perspective and, 37
  metaphors and, 167–170
  studies, 8
  token representations in, 40
  type representations and, 69
Causal connection, 118–119, *120*, 121, 132
  ad hoc subkinds and, 135
  agnosticism and, 129–130
  generalization from arbitrary instances and, 142
Causal explanations, 74
Characterizing properties, 111
  causal connections and, 118, 123, 132
  generics and prevalence of, 123–124
  statistical connections and, 118, 120, 132
  whole-part relations and, 122, 131
Class inclusion relations, 67
Coactivation packages, 149
Cognition
  arbitrary instances as tool of, 142
  Aristotle's categories and, 261–263
  core knowledge systems in, 205
  embodied or grounded approaches to, 8
  kind representations and, 255
Combinatorial thought, 205
Commonsense concepts, 3, 165, 243–244
  abstractness of perspectives from, 21–23
  essentialist approach and, 12–13
  generality of perspectives, 26
  generative system and, 213
  intricate and varied perspectives and, 23–25
  puzzles from perspectival nature of, 26–28, 31–33
  theory approach and, 10
  untyped representations and, 40
Community, 246–247

Complex concepts, 77, 233–234
Computational models
BPL framework and, 248
of conceptual representation, 9
Concept-dependent properties, 15–16
Concepts, 3, 265
ad hoc kind representation and, 103
in animals, 28
atomic, 269n3
classical theory of, 6
combination of, 135–136
complex, 77, 233–234
dual character, 93–94, 272n6
explanatory dimension of, 7
formal structures of, 8, 29
as generative programs, 248–250
sortal, 224
superordinate, 82, 84
theory approach to, 10–11
word meanings and, 14
Conceptual bootstrapping, 205–207
Conceptual empiricism, 218–221
Conceptual indices, 256–257
Conceptual knowledge, 71–73, 221, 240–241
brain damage and impairment of, 251
domain specificity of, in brain, 251–255
revision of, 271n7
Conceptual primitives, 214, 232–234
Conceptual relations, 150
individuation and polysemy patterns across, 160–162
individuation relation and, 165
Conceptual representation, 4–5, 51, 76
brain mechanisms underlying, 251–260
computational models of, 9
development and, 223
framework theories of, 10–11
investigating, in brain, 255–258
metaphor comprehension and, 168–170
for natural kinds, 74
nonperspectival approaches to, 6–9
perspectives and, 26–27
psychological essentialism and, 11–12
Concrete instance-of-kind representations, 97, 156
kinds intrinsically containing, 89–93, *91*
Constrained Variety of Conceptual Forms hypothesis, 80
Containers, 162
Content domains
explanatorily significant properties and, 98
explanatory structure and, 96
kind concept classes and, 239
lawful connections and, 57
properties connected across, 17
Contextual information
ad hoc instances and, 104–105
unit of quantification inference from, 155–156
Contextually determined individuation, 162–163
Core knowledge systems, 205–207
lexically expressible kind concepts difference from, 209–210
Counterexamples, 117
Cross-linguistic variation, 18–20

Definite singular generics, 137–139
Definite singular noun phrases, 137, 143
Definitions, 6
summary descriptions and, 43
Description. *See also* Summary descriptions
kind-determined modes of, 214, 229–232
Descriptive component, 255
Descriptive knowledge, 218–219
Descriptive universal generalization, 227
Development, conceptual representation and, 223
Distributed domain-specific hypothesis, 251–254
Distributional information, 198

Domain-general principles, 214, 238–240
Domain specificity, 214, 238–240
  of conceptual knowledge in brain, 251–255
Dual character concepts, 93–94, 272n6
Dual reference mechanism, 172–173, 275n1
Dyadic functional capacities, 253

Efficient explanations, 96–98, *97*
Embodied cognition, 8
Emergent perspectives approach, 193, 200–203
Emergent properties, in metaphor comprehension, 180
Essence, 121, 269n4
  formal explanation and, 120
Essentialist perspective, 11–12. *See also* Psychological essentialism
  assumptions in, 13
Exemplar approach, 7–9, 269n2, 270n2
Explanation
  categorization connection to, 22
  causal, 74
  distinct modes of, domains and, 95, *97*
  forms of, 96–97, *97*, 98, 273n8
  kind-determined modes of, 214, 229–232
  in kind representations, 96
  principled connections and, 273n8
  principle of, 227
  properties and, 73–74
  summary descriptions and, 43
  teleological, 74, 96–98, *97*
Explanatorily significant properties, 97, *97*, 98
Explanatory dimension, of concepts, 7
Explanatory knowledge, 218–219
Explanatory structure
  ad hoc subkinds and, 102
  classes of kinds distinguished by, 94–99, *97*, *99*
  domain differences in, 95–96
Explicit characterizations, of conceptual forms, 199
Extrinsic placeholders, 130

Factual knowledge, 223
Fairy tales, 227
Family, 246–247
Feature matching, 190
Features
  concept-dependent, 15–16
  descriptive, 93
  descriptive perceptual, 255
  perceptual, 255
  principled connections and abstract and concrete, 94
Ferrara, Dominick, 182
Fodor, Jerry, 232–233, 235, 237
Food, individuation of, 160–161
Formal aspect-kind relations, 131
Formal explanation
  classes of kinds distinguished by, 94–99, *97*, *99*
  defining, 96
  essence and, 120
  instance-of-kind perspective and, 41, 45–46
  properties and, 45, 94, 223, 230
  sets representations and, 49
  whole-part relations and, 116
  within-kind variation and, 127
Formal knowledge
  abstract, 140–141
  innate, 204
Formally distinct characters, *97*
Formal representations, 197
Formal structures
  of animate kind concepts, 253
  of concepts, 29
  instance-of-kind representations and, 63, *63*, 185
  of kind concepts, 52, 252
  of kind representations, 18, 57–58
  of lexically expressible concepts, 29, 73

metaphors and, 174–175
perspective encoded in, 193
of vehicle category for metaphor, 177
Frames, 33
of reference, 231
Framework theories, 10–11, 51
Functional capacities, 253
Fuzzy set theory, 49

Generality
of acquired principles, 222
of perspectives from commonsense concepts, 26
Generalizations
accidental, 139
descriptive universal, 227
forms of, 139–144
knowledge acquisition and, 218
in mathematics, 140–141
nonaccidental, 114–115, 139
quantitative, 110–111
representative mode of, 142–143
representing via conceptual form, 114–115
weak, 112, 131–133
General knowledge, parameter setting and acquisition of, 214, 218–222
Generative LLMs, 201–203
Generative mechanism, 104–105
ad hoc kinds and, 103
as component of kind representations, 258–260
for instance-of-kind representations, 55–56, *57*, 69, 257
kind concepts as, 113
modeling, 249
Generative programs
BPL framework and, 9, 248
concepts as, 248–250
Generative system
commonsense concepts and, 213
human conceptual system and, 203–204
perspectives and, 32
Generic knowledge, 109–110
acquisition of, 221
agnosticism and, 129
conceptual form of kind concepts and, 113–117
conceptual form of other types of, 117–121
nature of, 115–117
principled connections and, 117–118, 222
Generic language, 109–110
Generics
bare plural, 135–136
causal connections and, 118–119, 129–130, 132
characterizing property prevalence and, 123–124
conceptual form of kind concepts and, 114–115
conceptual forms and, 113–121, 144–145
definite singular, 137–139
grounding real-world conditions and, 113, 132
indefinite singular, 136–137
minority characteristic, 124–128
predicates and, 110–111
principled connections and, 201
puzzling phenomena concerning, 110–113
as rules, regulations, or laws, 111–113
semantics of, 112–113
statistical connection and, 118–121, 123, 130, 132
underspecification of conceptual form, 130
weak generalizations and, 131–133
Generic use of noun phrases, constraints on, 135–139
Gesture systems, 109
Goal-directed movement, 206, 253
Grammar, universal, 204, 236

Grounded cognition approaches, 8
Group-instance kinds, 246–247
Groups
perspectives for thinking and talking about, 244–247
structural explanations and, 263

Highly apt metaphors, 180–181
Homonyms, 147
Human beings
right-handedness and, 128–129
similarities and differences in instances of, 245–246

Ideal category members, 46
Identification via Combination approach, 216–218
Identity
of instance-of-kind representations, 256
principle of, 224–225, 227–228, 234
Indefinite singular generics, 136–137
Indefinite singular noun phrases, 142
Indexical component, 255
Indexical representation, 255–258
Indices
instance-of-kind representations and, 255–258
for kind representations, 157–158, 160
Indirect representation, 84
Individuated subkinds, 87, *88*, 89
Individuation
conceptually determined, 163–165
contextually determined, 162–163
infants and, 208
of instance-of-kind representations, 151, *152*, 153–157, 217–218
of kind representations, 151, *152*, 153, 159
across kinds and conceptual relations, 160–162
patterns of polysemy across kinds and, 157–160
principle of, 224, 226
of subkind-of-kind representations, 151, *152*, 153–154
systematic polysemy and, 165–166
Individuation for Polysemy principle, 150, 154, 157, 159–162, 165
Induction, 218
Inductive biases, 248
Infants
individuation and, 208
instance-of-kind representations and, 208
perspectives available to, 207–209
Inferential machinery, 133
Information
distributional, 198
newspapers and, 163–164
sortal, 228
Innate concepts, 235
Innate formal knowledge, 204
Innate knowledge system, 235–237
Innate principles, 222
Instance-of-kind exception, 185–186
Instance-of-kind perspective, 21–23, 32, 37, 243
abstract characteristics of, 38–42
ad hoc instances, 87, *88*, 89
approaches to representing kinds and, 42–50
formal explanation and, 41, 45–46
framework theories and, 51
identifying things and, 70
kind concepts and, 59–60
kind identity and, 70
member-of relation and, 44–45
normative expectations and, 41, 46–47
similarity and, 41–42, 47–48

summary description limitations and, 51
uniqueness and non-uniqueness relations and, 38–42
Instance-of-kind representations, 214, 267
abstract, 156, 158
concrete, 156
formal structures and, 63, *63*, 185
generative mechanism for, 55–56, *57*
individuation of, 158
infants and, 208
as intrinsic to kind representations, 216–218
kind concepts individuating, 151, *152*, 153–157
mechanisms for, *63*
metaphors and, 185–186
parts of, 56
perceptual index and, 256–257
subkinds and, 67
as unit of quantification, 153–155
untyped index and, 70–71
Instance properties, 90
Instances, 15
abstract, 89–93, *91*, 97
ad hoc, 87, *88*, 89, 105
in ad hoc subkinds, 100
arbitrary, 140–142
concrete, 89–93, *91*, 97
differences of, 53–54
existing actually or potentially, 60–61
individuation of, 217–218
of kinds, 69
kinds containing, 61–62
kinds containing abstract and concrete, 89–93, *91*
kinds intrinsically containing, 80, *81*, 84–87, *85*
member-of relation and, 44–45, 48
number of, for kinds, 52–53, 68
potentially existing, 103
principle of identity and, 225
representative, 142–143
simultaneous uniqueness and nonuniqueness with kind of, 43–44, 48
Intrinsic senses, 163–165
Intrinsic typing, 70
of particulars, 45, 48–49
"Is-it-one-or-two" tasks, 208

Jackendoff, Ray, 18

Kahlo, Frida, 161
Kahneman, Daniel, 274n2
Kind-determined frame of reference, 231
Kind-determined modes of explanation and description, 214, 229–232
Kind identity, 70–71, 84
Kind perspective, 21–23, 106
Kind predicates, 70
Kind representations
augmented mechanism for, 58, *59*
axes and frames of references and, 231
cognition and, 255
component representing character of kind in, 72
explanatory component of, 96
formal structure of, 18, 57–58
formation of, 220
framework theories and, 51
generative component of, 258–260
as generative mechanisms, 216
indices for, 157–158, 160
individuating, 151, *152*, 153, 159
instance-of-kind representations as intrinsic to, 216–218
metaphors and comparing, 184
principled connections and, 266

Kinds, 37
acquiring knowledge about, 218, 221
ad hoc, 102–105
approaches to representing, 42–50
artifact, 74
characterized and individuated in abstract and concrete terms, 93–94, *95*
class inclusion relations and, 67
components representing qualitative character of, 98
defining, 269n1
differences of instances of, 53
distinguished by formal explanatory structure, 94–99, *97*, *99*
existing actually, 63–64
existing potentially, 270n1
group-instance, 246–247
identifying nonaccidental connections to properties, 139–144
indexically distinguishing, 64
with individuated subkinds but ad hoc instances, 87, *88*, 89
individuation and polysemy patterns across, 160–162
instance-of-kind perspective and, 38–42
instances and subkinds contained in, 61–62
instances existing actually or not at all, 61
instances existing actually or potentially, 60–61
instances of, 15
intrinsically containing abstract and concrete instances, 89–93, *91*
intrinsically containing instances and no subkinds, 84–87, *85*
intrinsically containing instances and subkinds, 80, *81*
intrinsically containing subkinds but no instances, 82, *83*, 84
member-of relation and, 44–45, 48
minority characteristic generics and, 124
natural, 74
number of instances of, 52–53, 68
principled connection of properties to, 58, 72, 77, 94–95, 122
properties characteristic of, 47, 57, 69
qualitative character of, 54–55, 57, 266
representing, *63*, *65*
sets and, 68–69
simultaneous uniqueness and nonuniqueness within, 43–44
summary representations and, 69
superordinate, 82
ways of talking about, 133–144
well-established, 135, 137
whole-part relations with, 116, 122
Knowledge
abstract formal, 140–141
acquiring, about kinds of things, 218–219, 221
causal, 119
conceptual, 71–73, 221, 271n7
core systems for, 205
descriptive, 218–219
explanatory, 218–219
factual, 223
generic, 109–110, 113–121, 221–222
innate, 235–237
innate formal, 204
normative, 219
parameter setting and acquisition of, 214, 218–222
of particulars, 220
pretending and, 175
principles and, 72

Languages
abstract formal structure of, 19
diversity within, 19
generic, 109–110
thought and, 18–19
Large language models (LLMs), 201–203
Lawful connection, 31, 54–55, 57–58
generic knowledge and, 116–118, 120, 122–125
proportion of instances with properties and, 122
Lawful structures, 112
Laws
generics as, 111–113
weak generalizations and, 131–133
Learning
perspective acquisition and, 27
perspective identification problem in, 204
perspective selection cues, 28
word-learning biases, 32
Lenses
mental, 193, 197
metaphors as, 189–190
perspectives as, 213
Lexical distinctions, 19
Lexicalizing meanings, 18
Lexically expressible concepts, 4, 78
atomicity of, 267
conceptual form of, 30–31
core knowledge systems differences from, 209–210
formal structure of, 29, 73
generative system and, 32
mental lens in, 193
as perspectives, 213
as primitive but not innate, 214, 234–238
situating perspectives from, 33–34
Lexically expressible kind concepts, 113
bare plural generics and, 135
as conceptual primitives, 232–234
kind-determined modes of explanation and description and, 214, 229–232
polysemy patterns within conceptual form of, 150–157
principle of explanation and, 227
units of quantification and, 233
Linguistic cues, to conceptual form, 197–198
Linguistic evidence, 17
Literal attributions, 178–179
Living things, 11
domain specificity and, 239, 251
number of principled connections and, 95
LLMs. *See* Large language models

Macnamara, John, 224
Magazines, 16, 157–159, 163–165, 176, 196
Material explanations, 96–97, *97*, 98
Mathematics, 6
generalization in, 140–141
Meanings
concepts and word, 14
lexicalizing, 18
Member-of relation, 44–45, 48
Mental lens, 193, 197
Mental representations, 216
Merge (mechanism), 258–259
Metaphor comprehension
categorization-based approaches to, 168–170
conceptual mapping approach to, 170
conceptual representation and, 168–170
emergent properties in, 180
similarity-based approaches to, 168–169
Metaphorical attributions, 178

Metaphors, 167
abstract characteristics and, 170–171
category formation in understanding, 170–174
conceptual forms and, 174–177
constraints on interpretation of, 177
exceptional interpretation, 186
formal constraints on interpretation of, 183–189
formal structures and, 174–175
highly apt, 180–181
instance-of-kind exception and, 185–186
as lenses, 189–190
novel, 181–182
perspective identification view of, 177–183
principled connections and, 181
simile divergence from, 188–189
statistical connection and, 176, 179
unlike things compared and, 183–185
Minority characteristic generics, 124–129
Modal sensorimotor representations, 8–9
Monadic functional capacities, 253
Morphosyntactic form, 198

Natural kinds, 74, 119
Newspapers, 16, 148–149, 163–165, 196
Nonaccidental connection, 139–144
Nonaccidental generalizations
representing via conceptual form, 114–115
statistics and distinguishing, 139
Nondescriptive relations, conceptual forms and, 199–200
Normative connections, 221
Normative expectations, 37
instance-of-kind perspective and, 41, 46–47
kind concepts and, 60
principled connections and, 181
set representations and, 49–50
within-kind variation and, 127
Normative knowledge, 219
Noun phrases, 164
bare plural, 140
definite singular, 137, 143
generic use of, 135–139
indefinite singular, 142
information available from, 142
Novel metaphors, 181–182
Numerical differences, 61, 245
of instances of kinds, 53
qualitative difference and, 54

Object core knowledge system, 205–206
Object kind concepts, 99
basic-level, 87, 89, 208
Object kinds, 19–20
Objects
concepts and structure of, 15
semantic bootstrapping and, 204
syntactic, 258–259
Organizations, 176
individuation and, 159

Paradox of unlike things compared, 183–185
Parametric variation, 214, 238–240
Particulars
core knowledge system representations of, 213
intrinsic typing of, 45, 48–49
knowledge acquisition and, 218–219
knowledge of, 220
untyped, 44–45
Perception, 253
core knowledge systems and, 205
untyped representations in, 39
Perceptual dissimilarity, 92
Perceptual features, 255
Perceptual indices, 256–257
Personalities, 244–245
Persons, perspectives for thinking and talking about, 244–247

Perspective identification approach, 203–204
Perspective identification problem, 204
  conceptual bootstrapping and, 205–207
Perspective identification view, 177–180, 193
  emergent properties and, 180
  evidence for, 180–183
Perspectives, 265
  abstract, 21–23
  acquisition of, 27
  emergent, 193, 200–203
  generality of, 26
  generative system and, 32
  infants and availability of, 207–209
  intricate and varied, 23–25
  as lenses, 213
  lexically expressible concepts as, 213
  mapping, 27–28
  partially overlapping, 196
  puzzles raised by, of commonplace concepts, 26–28, 31–33
  representation of, 26–29
  selection cues for, 28
  situating, from lexically expressible concepts, 33–34
  syntactic and conceptual form and, 30–31
  for thinking and talking about persons and groups, 244–247
Picasso, Pablo, 161
Placeholders, 74, *75*, 120
  causal connections and, 119, 130
  statistical connections and, 130
Polysemous words, 147
Polysemy
  ad hoc instances and, 162–163
  ad hoc kinds and, 162–163
  conceptual form and patterns of systematic, 150
  conceptual structure and, 149
  contextually determined individuation and, 162–163
  forms of, 147–149
  individuation and patterns of, across kinds, 157–160
  lexically expressible kind concepts and patterns of, 150–157
  patterns of, across kinds and conceptual relations, 160–162
  systematic, 148–149, 160–162, 164–166, 175–176
Potentially existing instances, ad hoc kinds and, 103
Predicates
  generics and, 110–111
  kind, 70
  type representations as, 69
Pretending
  categories and, 175
  knowledge and, 175
  perspective and, 177
  properties and, 176
Principled connections, 58, 72, 77, *120*, 121, 266
  absence of property and, 127
  abstract and concrete features and, 94
  agnosticism and, 130
  attributions and, 181
  descriptive facts and, 221
  explanation and, 273n8
  generalization from arbitrary instances and, 142
  generic knowledge and, 117–118, 222
  generics and, 201
  grounding real-world conditions and, 132
  kind representations and, 266
  metaphors and, 181
  minority characteristic generics and, 124–125
  modeling, 249
  normative expectations and, 181
  number of, 95, 98, *99*

Principled connections (*continued*)
placeholders and, 120
of properties to ad hoc subkind, 102
of properties to kind, 94–95, 122
representative generalization and, 143
to vehicle category, 181–182
within-kind variation and, 127
Principled differences, 85, 89, 125, 269n5
between people, 245
Principled divisions, between types, 125
Principled within-kind variation hypothesis, 125–127
Principle of application, 224, 227
Principle of explanation, 227
Principle of identity, 224–225, 227–228, 234
Principle of individuation, 224, 226
Principles
acquisition of, 214, 222–224
domain-general, 214, 238–240
innate, 222
unrevisability of, 223
Properties
accidental, 185
attribution of, 178–179
categorization and explanation of, 40
causal connections and, 118–119
causal explanation and, 74
characteristic of kinds, 47, 57, 69
concept-dependent, 16
connecting across content domains, 17
distinguishing types of, 96
emergent, in metaphor comprehension, 180
explanatorily significant, 97, *97*, 98
formal explanation and, 45, 73, 94, 223, 230
identifying nonaccidental connections to kinds, 139–144
instance, 90
modeling, 249
modes of explanation and, *97*
number of, with principled connections, 95, 98, *99*
pretending and, 176
principled connection to ad hoc subkind of, 102
principled connection to kind of, 58, 72, 77, 94–95, 122
proportions of instances with, 122–129
representative generalization and, 143
semantic bootstrapping and, 204
specification of, 72
statistical connections and, 118–119
subkind differences and, 66
teleological explanation and, 74
whole-part relations with, 116, 122
Prototypes, 6–7, 9, 168
category representation and, 43, 46
Psychological essentialism, 11–12, 74, 230
assumptions in, 13
causal understanding and, 119
essence, 121, 269n4
formal explanation and, 120
placeholders and, 74–75, 120

Qualitative character, 126
components representing, 98
Qualitative differences
between instances, 56, 89–90
instances of kinds and, 61
of kinds, 54, 57
subkinds and, 66
Quantification
instance-of-kind representations as units of, 153–155
lexically expressible kind concepts specifying units of, 233
subkind-of-kind representations as units of, 153–154
Quantitative generalizations, 110–112, 122

Reasoning
  conceptual bootstrapping and, 205
  errors and fallacies in, 274n2
  untyped representations and, 39–40
Reference frames, 231
Regulations, generics as, 111–113
Reicher, Cameron, 181
Representation, 270n3. *See also* Conceptual representation; Instance-of-kind representations; Kind representations
  of ad hoc kinds, 103
  of categories as sets, 48–50
  indexical, 255–258
  indirect, 84
  instance-of-kind, 55–56, *57*
  of kind concepts, 52–68
  of kinds, *63*, *65*
  of minority characteristic generics, 124–128
  of perspectives, 26–29
  of subcategories, 67–68
  untyped, 39
Representative mode of generalization, 142–143
Representativeness heuristic, 274n2
Ritchie, Katherine, 150
Rivera, Shamauri, 180
Roles, 247
Rules, generics as, 111–113

Sapir, Edward, 18
Selective impairment, 251
Semantic bootstrapping theories, 204, 207
Semantics
  as agnostic system, 113, 129, 131
  of generics, 112–113
Sensorimotor primitives, 235
Sentence interpretation, syntactic structure and, 29
Sets
  categories as, 48–50
  kinds and, 68–69
Set-theoretic approaches, 69
Sign language, 109
Similarity
  instance-of-kind perspective and, 41–42, 47–48
  metaphor comprehension and, 168–169
  relevance without dependence, 47–48, 50
Similes, 183, 188–189
Social kind concepts, 119
Sortal concepts, 224, 275n2
Sortal information, 228
Sortal principles, 177, 214, 224–229
Statistical connection, 118–121, *120*, 123, 132, 201
  ad hoc subkinds and, 135
  agnosticism and, 130
  generalization from arbitrary instances and, 142
  metaphors and, 176, 179
Structural explanations, 263
Structure matching, 190
Subcategories, 86
  representation of, 67–68
Subkind-of-kind representations
  kind concepts individuating, 151, *152*, 153–154
  as unit of quantification, 153–154
Subkinds, 61–62, 64
  abstract, 89
  ad hoc, 99–102, *101*, 105, 135
  class inclusion relations and, 67
  differences between, 66
  existing actually, 66
  individuated, 87, *88*, 89
  instance-of-kind representations from, 67
  kinds intrinsically containing, 80, *81*, 82, *83*, 84
  limiting membership of, 102
  properties and, 66
Subsets, minority characteristic generics and, 124–125

Substance kind concepts, 99, 206
Substance kinds, 19–20
Summary descriptions, 6–7, 51
  definitions and, 43
  explanation and, 43
  limitations of, 51
  member-of relation and, 44–45
  nonuniqueness and, 44
  particulars and, 45
  representing categories and, 42–48
Summary representations, 168, 270n2
  kinds and, 69
Superordinate concepts, 82, 84, 138
Syntactic bootstrapping, 272n7
Syntactic form
  conceptual form differences from, 31
  perspectival properties and, 30–31
Syntactic objects, 258–259
Syntactic structure
  perspectival properties of, 30–31
  sentence interpretation and, 29
Systematic polysemy, 148–149, 160–162, 164–166, 175–176
  conceptual form and patterns of, 150

Teleological explanations, 74, 96–97, *97*, 98
Thought
  combinatorial, 205
  languages and, 18–19
  systems of, 205
  untyped representations and, 39–40
Token representations, 40, 70, 216–217
Topic category, 181
Topic concept, 177
Topic kind, 187
Triangles, 140–141, *141*
Tversky, Amos, 274n2
Type representations, 216. *See also* Intrinsic typing
  categorization and, 69
Types
  modeling, 249
  principled divisions between, 125
Type-token distinction, 249

Uniform Conceptual Form hypothesis, 79, 104, 272n1
Units of quantification
  inference from contextual information of, 155–156
  instance-of-kind representations as, 153–155
  lexically expressible kind concepts and, 233
  subkind-of-kind representations as, 153–154
Universal grammar, 204, 236
Unlike things, paradox of comparing, 183–185
Untyped index, 70–71
  ad hoc subkinds and, 101
Untyped particulars, 44–45
Untyped representations, 39–40
Utterances, generic, 109

Vehicle category, 169–173, 176–177, 181–182, 187
Vehicle concept, 176–177, 179–180
  conceptual form of, 180
Visual working memory, 255–257

Weak generalizations, generating, 131–133
Well-established kinds, 135, 137
Whole-part relations, 116, 122, 131
Whorf, Benjamin, 18
Word-learning biases, 32
Word meanings, concepts and, 14
Working memory
  untyped representations in, 39
  visual, 255–257